THE
Explosive Child

A New Approach
for Understanding and Parenting
Easily Frustrated,
"Chronically Inflexible" Children

Ross W. Greene, Ph.D.

HarperCollins*Publishers*

In memory of Irving A. Greene

HarperCollins books may be purchased for educational, business, or sales promotional use. For information please write: Special Markets Department, HarperCollins Publishers, Inc., 10 East 53rd Street, New York, NY 10022.

FIRST EDITION

Designed by Elina D. Nudelman

Library of Congress Cataloging-in-Publication Data

Greene, Ross W.
The explosive child : a new approach for understanding and parenting easily frustrated, "chronically inflexible" children / Ross W. Greene.—1st ed.
p. cm.
ISBN 0-06-017534-6
1. Problem children. 2. Behavior disorders in children. 3. Child rearing. 4. Parent and child. I. Title.
HQ773.G73 1998
649'.153—dc21 98-23753

00 01 02 03 04 ❖/RRD 20 19 18 17

Anyone can become angry, that is easy . . .
but to be angry with the right person, to the right degree,
at the right time, for the right purpose, and in the right
way . . . this is not easy.

—*Aristotle*

Contents

Acknowledgments

This book is about helping inflexible, easily frustrated, explosive children and their parents think and interact more adaptively. My thinking has been influenced by two of the world's preeminent clinician-researchers in child psychopathology, Drs. Thomas Ollendick and Joseph Biederman. It has been my incredible good fortune to work with these distinguished scholars. Tom, my mentor while I was in graduate school at Virginia Tech, taught me how to think critically; Joe, my colleague at Massachusetts General Hospital, has furthered the process. Although both would appreciate the need to subject much of what is written in this book to rigorous scientific evaluation, I suspect they would be reassured by the belief that children and their families may be helped by its contents.

Though I have benefited from my interactions with many mental health and education professionals over the years, two psychologists who supervised me during my training were particularly influential: Dr. George Clum at Virginia Tech and Dr. MaryAnn McCabe at Children's National Medical Center in Washington, DC. I've also found much wisdom in the writings of several authors, including Dr. Robert Brooks (author of *The Self-Esteem Teacher*); Dr. Stanley Turecki (author of

The Difficult Child); Dr. Thomas Gordon (author of *P.E.T.: Parent Effectiveness Training*); and Dr. Myrna Shure (author of *Raising a Thinking Child*). And I probably wouldn't have gone into psychology in the first place if I hadn't stumbled across the path of Dr. Elizabeth Altmaier when I was an undergraduate at the University of Florida.

Love helps, too, and I get lots from my wife, Melissa, daughter, Talia, mother, brother, sister, aunt, and grandmothers. I also got lots from those whose deaths preceded the publication of this book: my father (to whom this book is dedicated); grandfathers; and, of course, Nene and Lucy Greene.

My agent, Wendy Lipkind, believed in this book the minute she saw the first draft. So did Joëlle Delbourgo at HarperCollins. Their collective vision for *The Explosive Child* has been critical to its development. The patience and superb editing skills of Tim Duggan, also at HarperCollins, have been indispensable as well.

Needless to say, those who were most central to the evolution of many of the ideas in this book and to whom I owe the greatest debt of gratitude were the many parents who entrusted me with the care of their children and families.

Although there are many inflexible-explosive girls, for ease of exposition most of this book is written in the masculine gender. The names and identifying information of all the children are fictitious. Most of the names are those of friends, nonspecific cartoon characters, rock musicians, and my wife's ex-boyfriends. Any resemblance to actual children of the same names is, as the saying goes, purely coincidental.

Preface

As the title suggests, this book is about explosive, inflexible, easily frustrated children. These children often exhibit severe behaviors—intense temper outbursts, noncompliance, volatility, mood instability, and verbal and physical aggression—that have the potential to make life extraordinarily challenging and frustrating for them, their parents, siblings, teachers, and others who interact with them. Though they have been described in many ways—difficult, strong-willed, coercive, manipulative, attention seeking, willful, contrary, and intransigent—and may carry any or many of various psychiatric diagnoses, such as oppositional-defiant disorder (ODD), attention-deficit/hyperactivity disorder (ADHD), Tourette's disorder, depression, bipolar disorder, and obsessive-compulsive disorder (OCD), I believe their behavior is still poorly understood and therefore difficult to change.

For a long time, the prevailing view of such behavior has been that it is the by-product of inept parenting practices. But research in the past ten to fifteen years has suggested to me that inflexible and explosive behavior is a lot more complex than has been previously thought and may emanate from a variety of different factors. Thus, there is no "one size fits all" approach to helping these children.

In writing *The Explosive Child,* my intent was to provide a more up-to-date conceptualization of such behavior and to describe a new, practical, comprehensive approach aimed at decreasing adversarial interactions between parents and children, reducing family hostility, and improving children's capacities for flexibility, frustration tolerance, communication, and self-regulation. My goals are to help you appreciate the manner in which your explanations for and interpretations of a child's inflexible-explosive behavior can influence how you respond to it, to help you understand the neurobiochemical factors that may lead to inflexible-explosive behavior in children, and to highlight the importance of finding the right treatment to fit the needs of individual children and families. Most of all, my goal is to help you break down the barriers imposed by your child's inflexibility and explosiveness and begin to interact in ways that feel better to you both.

This approach has evolved over the eighteen years that have elapsed since I worked with my first inflexible-explosive child. I don't think the children have changed all that much since then, but my approach to helping them and their parents and teachers is a lot different. And I think it works a lot better.

The only prerequisite is an open mind.

Ross W. Greene, Ph.D.
Boston, Massachusetts

1
The Waffle Episode

Jennifer, age eleven, wakes up, makes her bed, looks around her room to make sure everything is in its place, and heads into the kitchen to make herself breakfast. She peers into the freezer, removes the container of frozen waffles, and counts six waffles. Thinking to herself, "I'll have three waffles this morning and three tomorrow morning," Jennifer toasts three waffles and sits down to eat.

Moments later, her mother and five-year-old brother, Adam, enter the kitchen, and the mother asks Adam what he'd like to eat for breakfast. Adam responds, "Waffles," and the mother reaches into the freezer for the waffles. Jennifer, who has been listening intently, explodes.

"He can't have the frozen waffles!" Jennifer screams, her face suddenly reddening.

"Why not?" asks the mother, her voice and pulse rising, at a loss for an explanation of Jennifer's behavior.

"I was going to have those waffles tomorrow morning!" Jennifer screams, jumping out of her chair.

"I'm not telling your brother he can't have waffles!" the mother yells back.

"He can't have them!" screams Jennifer, now face-to-face with her mother.

The mother, wary of the physical and verbal aggression of which her daughter is capable during these moments, desperately asks Adam if there's something else he would consider eating.

"I want waffles," whimpers Adam, cowering behind his mother.

Jennifer, her frustration and agitation at a peak, pushes her mother out of the way, seizes the container of frozen waffles, then slams the freezer door shut, pushes over a kitchen chair, grabs her plate of toasted waffles, and stalks to her room. Her brother and mother begin to cry.

Jennifer's family members have endured literally thousands of such episodes. In many instances, the episodes are more prolonged and intense and involve more physical or verbal aggression than the one just described (when Jennifer was eight, she kicked out the front windshield of the family car). Mental health professionals have told Jennifer's parents that Jennifer has something called oppositional-defiant disorder. For the parents, however, a simple label doesn't begin to explain the upheaval, turmoil, and trauma that Jennifer's outbursts cause. Her siblings and mother are scared of her. Her extreme volatility and inflexibility require constant vigilance and enormous energy from her mother and father, thereby lessening the attention the parents wish they could devote to Jennifer's brother and sister. Her parents frequently argue over the best way to handle her behavior, but agree about the severe strains Jennifer places on their marriage. Although she is above average in intelligence, Jennifer has no close friends; children who initially befriend her eventually find her rigid personality difficult to tolerate.

Over the years, the parents have sought help from countless mental health professionals, most of whom advised them to set firmer limits and to be more consistent in managing Jennifer's behavior, and instructed them how to implement formal behavior management strategies. When such strategies failed to work, Jennifer was medicated with innumerable combinations of drugs, without dramatic effect. After eight years of medicine, advice, sticker charts, time-outs, and reward programs, Jennifer has changed little since her parents first noticed there was something "different" about her when she was a toddler.

"Most people can't imagine how humiliating it is to be scared of your own daughter," Jennifer's mother once said. "People who don't have a child like Jennifer don't have a *clue* about what it's like to live like this. Believe me, this is not what I envisioned when I dreamed of having children. This is a nightmare."

"You can't imagine the embarrassment of having Jennifer 'lose it' around people who don't know her," her mother continued. "I feel like telling them, 'I have two kids at home who don't act like this—I really *am* a good parent!'"

"I know people are thinking, 'What wimpy parents she must have . . . what that kid really needs is a good thrashing.' Believe me, we've tried everything with her. But nobody's been able to tell us how to help her . . . no one's really been able to tell us what's the *matter* with her!"

"I hate what I've become. I used to think of myself as a kind, patient, sympathetic person. But Jennifer has caused me to act in ways I never thought I was capable of. I'm emotionally spent. I can't keep living like this.

"I know a lot of other parents who have pretty difficult children . . . you know, kids who are hyperactive or having trouble paying attention. I would give my left arm for a kid who was just hyperactive or having trouble paying attention! Jennifer is in a completely different league! It makes me feel very alone."

The truth is, Jennifer's mother is not alone; there are a lot of Jennifers out there. Their parents quickly discover that strategies that are usually effective for shaping the behavior of other children—such as explaining, reasoning, reassuring, nurturing, redirecting, ignoring, rewarding, and punishing—don't have the same success with their Jennifers. Even formal behavior management programs—sticker charts, contingent rewards and punishments, and time-outs—and commonly prescribed medications have not led to satisfactory improvement. If you started reading this book because you have a Jennifer of your own, you're probably familiar with how frustrated, confused, angry, bitter, guilty, overwhelmed, worn out, and hopeless Jennifer's parents feel.

Besides oppositional-defiant disorder, children like Jennifer may be diagnosed with any of a variety of psychiatric disorders and learning inefficiencies, including attention-deficit/hyperactivity disorder (ADHD), mood disorders (bipolar disorder and depression), Tourette's disorder, anxiety disorders (especially obsessive-compulsive disorder), language-processing impairments, sensory integration deficits, nonverbal learning disabilities, and even Asperger's disorder. Such children may also be described as having difficult temperaments. Whatever the label, children like Jennifer are distinguished by a few characteristics—namely, striking inflexibility and low frustration tolerance—that make life significantly more difficult and challenging for them and for the people who interact with them. These children often seem unable to shift gears and think clearly in the midst of frustration and respond to even simple changes and requests with extreme inflexibility and often verbal or physical aggression. For reasons that will become clear, I've come to refer to such children as *inflexible-explosive*—not because I'm interested in inventing yet another syndrome (there are plenty already)—but because I think it's more accurate and descriptive and therefore gives a better idea about what they need help with.

Whatever the label, children like Jennifer are distinguished by a few characteristics—namely, striking inflexibility and low frustration tolerance—that make life significantly more difficult and challenging for them and for the people who interact with them.

How are inflexible-explosive children different from other children? Let's take a look at how different children may respond to a fairly common family scenario. Imagine that Child 1—Hubert— is watching television and his mother asks him to set the table for dinner. Hubert has a pretty easy time shifting from his agenda—watching television—to his mother's agenda—setting the table for dinner. Thus, in response to, "Hubert, I'd like you to turn off the television and come set the table for dinner," he would likely reply, "OK, Mom, I'm coming" and would set the table shortly thereafter.

Child 2—Jermaine—is somewhat tougher. He has a harder time shifting from his agenda to his mother's agenda but is able to manage his frustration and shift gears (often with a threat hanging over his head). Thus, in response to, "Jermaine, I'd like you to turn off the television and come set the table for dinner," Jermaine may initially shout, "*No way*, I don't want to right now!" or complain, "You *always* ask me to do things right when I'm in the middle of something I like!" However, with some extra help (Mother: "Jermaine, if you don't turn off the television and come set the dinner table right now, you're going to have to take a time-out"), these "somewhat tougher" children do shift gears.

And then there is Jennifer, Child 3, the inflexible-explosive child, for whom shifting gears—from her agenda to her mother's agenda—often induces an unimaginable, intense, debilitating level of frustration. In response to, "Jennifer, I'd like you to turn off the television and come set the table for dinner," these children get stuck and often simply explode, at which point all bets are off on what they may say or do.

Inflexible-explosive children come in all shapes and sizes. Some blow up literally dozens of times every day; others do so only a few times a week. Many "lose it" only at home, others only at school, some both at home and school. Some scream when they become frustrated but do not swear or become physically or verbally aggressive. One such child, Richard, a spunky, charismatic fourteen year old who was diagnosed with ADHD, began to cry in our first session when I asked if he thought it might be a good idea for us to help him start managing his frustration so he could begin getting along better with his family members. Others scream and swear but do not lash out physically (including Jack, an engaging, smart, moody ten year old, diagnosed with ADHD and Tourette's disorder, who had a reliable pattern of becoming inflexible and irrational over the most trivial matters and whose inflexibility and irrationality tended to elicit similar behaviors from his parents). Still others combine the whole package, such as Marvin, a bright, active, impulsive, edgy, easily agitated eight year old with Tourette's disorder, depression, and ADHD, who reacted to unexpected changes with unimaginable screaming, swearing, and physical violence (on one occasion, Marvin's father innocently turned off an unnecessary light in the room in which Marvin was playing a video game, prompting a massive one-hour blowup).

What should become quite clear as you read this book is that these children have wonderful qualities and tremendous potential. In most ways, their general cognitive skills have developed at a normal pace. Yet their inflexibility and low tolerance for frustration often obscure their more positive traits and cause them and those around them enormous pain. I can think of no other group of children who are so misunderstood. Their parents are typically caring, well-intentioned people who often feel guilty that they are no longer able to feel great love for their children.

"You know," Jennifer's mother would say, "each time I start to get my hopes up . . . each time I have a pleasant interaction with Jennifer . . . I let myself become a little optimistic and start to like her again . . . and then it all comes crashing down with her next explosion. I'm ashamed to say it, but a lot of the time I really don't like her, and I definitely don't like what she's doing to our family. We are in a perpetual state of crisis."

Clearly, there's something different about the Jennifers of the world. This is a critical, often painful, realization for parents to come to. But there is hope, as long as their parents, teachers, relatives, and therapists are able to come to grips with a second realization: Inflexible-explosive children often require different disciplinary practices than do other children. Unfortunately, there is no bible on how to deal with these children, particularly if medication and standard behavior management strategies fail to resolve their difficulties. Thus, the parents, teachers, relatives, and therapists of such children often aren't sure what to do or where to turn.

What I've found is that dealing more effectively with inflexibility-explosiveness requires, first and foremost, a new *understanding* of what these children are about. Once parents have a better sense of why these children behave as they do, strategies for helping things improve become clearer. In some instances, helping parents achieve a more accurate understanding of their child's difficulties can, by itself, lead to improvements in parent-child interactions, even before formal strategies are tried. The first chapters of this book are devoted to helping you think about why these children adapt so poorly to changes and requests, are so easily frustrated, and explode so quickly and so often. At the same time, you'll read about why popular strategies for dealing with them may be less effective than expected. In later chapters, you'll read about alternative strategies that have been helpful to many of the children and families with whom I've worked over the years.

If you are the parent of an inflexible-explosive child, this book may restore some sanity and optimism to your family and help you feel that you can actually handle your child's difficulties confidently and competently. If you are a relative, friend, teacher, or therapist, this book should, at the least, help you understand. There is no panacea. But there is hope.

2
Terrible Beyond Two

One of the most amazing and gratifying things about being a parent is watching your child develop new skills and master increasingly complex tasks with each passing year. Crawling progresses to walking and then advances to running; babbling slowly develops into full-blown talking; smiling progresses to more sophisticated forms of socialization; learning the letters of the alphabet leads to reading whole words and then sentences, paragraphs, and books.

It is also amazing how unevenly children's different skills develop. Some children learn to read more readily than they learn to do multiplication. Some children turn out to be excellent athletes, while others are less athletically skilled. When children's skills in a particular area lag well behind their expected development, we often give them special help. This special help may be informal, such as Steve's baseball coach suggesting extra batting practice, or formal, such as Ken's school giving him remedial assistance with reading. In some cases, skills may lag because of the children's lack of exposure to the appropriate material (for example, maybe Steve was never actually taught how to hit a baseball). More commonly, children have difficulty learning a particular skill even though they have been provided with the instruction typically needed to master the skill and have the desire to master it. It's not that they don't *want* to learn; it is simply that they are not learning as readily as expected. Sometimes, with intensive instruction, extra effort, and a lot of support and encouragement, children can improve their skills in areas in which they are having difficulty. And sometimes children's skills improve to some extent but never reach the desired level of mastery. In other words, some children—no matter how hard they and their parents, teachers, and coaches try—are just not "built" to be great athletes, readers, or spellers.

Flexibility and *frustration tolerance* are also skills we expect to develop in children as they grow past the toddler years. Most children go through a phase called the Terrible Twos, in which they are increasingly inflexible and easily frustrated. During this phase, there is often a marked increase in the frequency and intensity of a child's tantrums; indeed, at this stage of development, tantrums are a common, if not particularly pleasant, way for children to say things like, "I'm hungry," "I'm tired," "I'm confused," or "I'm not happy with what's going on at the moment." To the tremendous relief of parents, this phase is usually relatively short-lived.

Why do most children eventually progress beyond the Terrible Twos? They do so because they gradually develop the ability to express their needs and desires more effectively, tolerate frustration and delay gratification more readily, "shift gears" more efficiently when events vary from the original plan, resolve problems by thinking through possible solutions, and move beyond "black-and-white" thinking by demonstrating the ability to appreciate the "gray" in many situations. In most cases, children need a little initial guidance from adults to recognize the necessity of these skills ("George, don't *hit* Carolyn! Use your *words*!" "Fred, if you won't let Cindy take a turn at the Nintendo, then I'm going to turn the whole thing off and neither of you will get to play." "Tom, I don't think Mary is enjoying watching you play with the Legos. Why don't you ask her if there's something she'd like to build.")

Mastery of these hurdles is crucial to a child's overall development because interacting adaptively with the world requires the continuous ability to resolve problems and effectively handle frustrations. Indeed, it's difficult to imagine many situations in a child's day that don't require flexibility, adaptability, and frustration tolerance. When two children disagree about which game to play, we hope both children have compromising and problem-solving skills that will help them resolve the dispute in an agreeable, mutually satisfactory manner. When bad weather forces parents to cancel their child's much-anticipated trip to the amusement park, we hope the child has the ability to express his disappointment appropriately, shift gears, and delay the trip and settle on an alternative plan. When a child is engrossed in a video game and it's time to set the table for dinner, we hope the child is able to interrupt his game, modulate his natural feelings of frustration, and think clearly enough to recognize that he can return to the game later. And when a child decides she'll have three frozen waffles for breakfast today and three tomorrow and her younger brother decides he wants three frozen waffles today, too, we hope the child can move beyond black-and-white thinking ("I am *definitely* going to have those three waffles for breakfast tomorrow, so there's no way my brother can have them") and recognize the gray in the situation ("I guess I don't have to eat those *exact* waffles. . . . I can ask my mom to buy more. . . . Anyway, I might not even feel like eating waffles tomorrow").

Just as some children lag in acquiring reading skills and other children do not develop great athletic skills, still others—the children this book is about—do not progress to the degree we would have hoped in the skills of flexibility and frustration tolerance. Thus, a major premise of this book is that these children do not *choose* to be explosive and noncompliant—any more than a child would *choose* to have a reading disability—but are delayed in the process of developing the skills that are critical to being flexible and tolerating frustration or have great difficulty applying these skills when they most need to. A second important premise of this book is that the inflexibility and explosiveness of these children have a lot to do with how they're "built."

When people say a child is "built" to be one way or another, they are often referring to the workings of the child's brain. Our understanding of these "workings" is still fairly elementary, though we live in an age in which science is quickly unraveling the chemical and physiological bases for human behavior. As we learn more, many popular explanations for behavior become obsolete. Old explanations for disorders, such as schizophrenia and depression, have fallen by the wayside as convincing evidence has emerged that in the preponderance of cases, there is a neurobiochemical predisposition to these disorders that is often genetically transmitted.

A major premise of this book is that these children do not choose to be explosive and noncompliant—any more than a child would choose to have a reading disability—but are delayed in the process of developing the skills that are critical to being flexible and tolerating frustration or have great difficulty applying these skills when they most need to.

Nevertheless, the human brain is a remarkably complex entity, so it's hard to get a precise handle on things, especially in children, whose brains aren't fully developed. Science has not advanced to the degree that one can speak with perfect accuracy about the specific neurobiochemical brain mechanisms underlying many childhood psychiatric disorders. Still, there's some fairly compelling evidence to suggest that, for example, irregularities in the prefrontal and frontal regions of the brain and other brain structures connected to these regions may contribute to the impairments in "executive" thinking skills often seen in children with ADHD. As you'll read, these skills—including mental flexibility, the ability to shift from one mind-set to another, problem solving, planning, organizing one's thoughts, and controlling one's impulses—are critical to one's capacity for flexibility and frustration tolerance. Other brain irregularities involving the neurotransmitter serotonin have been implicated in depression, aggression, and obsessive-compulsive disorder, all of which can be associated with inflexibility and explosiveness. Still other irregularities—largely in the right hemisphere of the brain—are thought to underlie impaired "nonverbal skills." Children who manifest what is known as nonverbal learning disability often have poor problem-solving skills, difficulty understanding cause-and-effect relationships; trouble adapting to unfamiliar, unexpected, or complex situations; and deficits in social perception and social judgment. As you'd imagine, these difficulties also have the potential to contribute to inflexible-explosive behavior. Finally, irregularities in certain regions of the left hemisphere of the brain may affect language functioning. Children with impaired language-processing skills often have difficulty sorting through and expressing their thoughts so as to solve problems and handle frustration adaptively.

One of the nice things about uncovering these physiological connections—imprecise as some still are—is that they help move us away from the standard, knee-jerk explanations for why some children keep doing things we wish they wouldn't; for instance,

"He's doing it for attention" or "He could do better if he really wanted to." It's becoming less common these days to hear someone suggest, for example, that a child with ADHD is "intentionally" being hyperactive, impulsive, and inattentive or that the vocal tics of a child with Tourette's disorder are "for attention." Of course, we are left to ponder why such knee-jerk explanations are still commonly invoked to explain the inflexibility-explosiveness that often accompanies these otherwise "unintentional" disorders.

The main point is this: There's a big difference between viewing inflexible-explosive behaviors as the result of a brain-based failure to progress developmentally and viewing them as planned, intentional, and purposeful. That's because your interpretation of a child's inflexible-explosive behaviors will be closely linked to how you try to change these behaviors. In other words, **your interpretation will guide your intervention**.

If you interpret a child's behavior as planned, intentional, and purposeful, then labels such as "stubborn," "manipulative," "coercive," "bratty," "attention seeking," "controlling," "resistant," and "defiant" will sound perfectly reasonable to you, and popular strategies aimed at motivating compliant behavior and "teaching the child who's boss" will make perfect sense. If this has been your interpretation of your child's explosive behavior, you're not alone. You're also not alone if this interpretation hasn't led you to a productive outcome. Throughout this book, I encourage you to put this motivational explanation on the shelf and give some consideration to the alternative explanation: that your child's explosive behavior is unplanned and unintentional and reflects a physiologically based developmental delay in the skills of flexibility and frustration tolerance. From this perspective, putting a lot of energy into motivating your child and teaching him who's boss may actually be counterproductive.

Indeed, these goals can actually fuel an adversarial pattern that may make progress much more difficult to achieve.

Is this alternative, contemporary view more accurate in many instances? If so, is there a better way of describing these children? And are there alternative strategies that may better match the needs of inflexible-explosive children and their families?

Yes, yes, and yes.

To begin with, here's the diagnostic criteria for oppositional-defiant disorder, which, as you read in Chapter 1, is the standard diagnosis for children whose temper outbursts, excessive arguing, and defiance persist well beyond the Terrible Twos:

Partial Diagnostic Criteria for Oppositional-Defiant Disorder*

A pattern of negativistic, hostile, and defiant behavior lasting at least six months, during which four (or more) of the following are present:

- often loses temper
- often argues with adults
- often actively defies or refuses to comply with adults' requests or rules
- often deliberately annoys other people
- often blames others for his or her mistakes or misbehavior
- is often touchy or easily annoyed by others
- is often angry or resentful
- is often spiteful and vindictive

*Reprinted with permission from the *Diagnostic and Statistical Manual of Mental Disorders,* Fourth Edition. Copyright 1994 American Psychiatric Association.

I'm not invested in these criteria because I don't think they do justice to the actual processes of inflexible-explosive behavior. These criteria also imply that oppositional behavior is deliberate, and I don't think this is true in many instances. If my first goal is to help you come to a more accurate understanding of your child's behavior, I'd better give you a more accurate way of describing it. The characteristics discussed next may be helpful:

Common Characteristics of Inflexible-Explosive Children

1 A remarkably limited capacity for flexibility and adaptability and incoherence in the midst of severe frustration. The child often seems unable to shift gears in response to parents' commands or a change in plans and becomes quickly overwhelmed when a situation calls for flexibility and adaptability. As the child becomes frustrated, his or her ability to "think through" ways of resolving frustrating situations in a manner that is mutually satisfactory becomes greatly diminished; the child has difficulty remembering previous learning about how to handle frustration and recalling the consequences of previous inflexible-explosive episodes, has trouble thinking rationally, may not be responsive to reasoned attempts to restore coherence, and may deteriorate even further in response to punishment.

2 An extremely low frustration threshold. The child becomes frustrated far more easily and by far more seemingly trivial events than other children of his or her age.

Therefore, the child experiences the world as one filled with frustration and uncomprehending adults.

3 An extremely low tolerance for frustration. The child is not only more easily frustrated, but experiences the emotions associated with frustration more intensely and tolerates them far less adaptively than do other children of the same age. In response to frustration, the child becomes extremely agitated, disorganized, and verbally or physically aggressive.

4 The tendency to think in a concrete, rigid, black-and-white manner. The child does not recognize the gray in many situations ("Mrs. Robinson is *always* mean! I *hate* her!" rather than "Mrs. Robinson is usually nice, but she was in a really bad mood today"); may apply oversimplified, rigid, inflexible rules to complex situations; and may impulsively revert to such rules even when they are obviously inappropriate ("We *always* go out for recess at 10:30. I *don't care* if there's an assembly today, *I'm* going out for recess!").

5 The persistence of inflexibility and poor response to frustration despite a high level of intrinsic or extrinsic motivation. The child continues to exhibit frequent, intense, and lengthy meltdowns even in the face of salient, potent consequences.

6 Inflexible episodes may have an out-of-the-blue quality. The child may seem to be in a good mood, then fall

apart unexpectedly in the face of frustrating circumstances, no matter how trivial.

7 The child may have one or several issues about which he or she is especially inflexible—for example, the way clothing looks or feels, the way food tastes or smells, and the order or manner in which things must be done.

8 The child's inflexibility and difficulty responding to frustration in an adaptive manner may be fueled by behaviors—moodiness/irritability, hyperactivity/impulsivity, anxiety, obsessiveness, social impairment—commonly associated with other disorders.

9 While other children are apt to become more irritable when tired or hungry, inflexible-explosive children may completely fall apart under such conditions.

What I hope these unofficial characteristics help you begin to do is understand and describe your child's explosions and inflexibility differently. It's hard to imagine how a child could be actively yanking your chain or know just the right buttons to push when he's not thinking coherently in the midst of frustration. It's harder still to imagine why a child would intentionally behave in a way that makes other people respond in a manner that makes him miserable. I also don't think these children are especially angry, though I do think they're extremely frustrated. If the term *angry* applies to them, it's because they're angry at being misunderstood. They typically don't understand their own behavior, but they're quite certain no one else does either. (Some children have been abused and neglected and have legitimate reasons to be angry and hostile or depressed. Most of the inflexible-explosive children with whom I've worked have neither been abused nor neglected.)

In their own incredible frustration and confusion over the constant inflexibility and explosiveness being exhibited by their child, parents frequently demand that the child provide a logical explanation for his actions. The dialogue often goes something like this:

It's hard to imagine how a child could be actively yanking your chain or know just the right buttons to push when he's not thinking coherently in the midst of frustration. It's harder still to imagine why a child would intentionally behave in a way that makes other people respond in a manner that makes him miserable.

Parent: We've talked about this a million times. . . . WHY DON'T YOU DO WHAT YOU'RE TOLD? WHAT ARE YOU SO ANGRY ABOUT?

Inflexible-explosive child: I don't know.

The child's maddening response usually has the effect of further heightening his parents' frustration. It's worth noting, of course, that the child is probably telling the truth.

In a perfect world, the child would respond with something like, "See, Mom and Dad, I have this little problem. You guys—and lots of other people—are constantly asking me to shift from Agenda A to Agenda B, and I'm not very good at it. In fact, when you ask me to make these shifts, I start to get frustrated. And when I start getting frustrated, I have trouble thinking clearly and then I get even more frustrated. Then you guys get mad. Then I start doing things I wish I didn't do and saying things I wish I didn't say. Then you guys get even madder and punish me, and it gets really messy. After the dust settles—you know, when I start thinking clearly again—I end up being really sorry for the things I did and said. I know this isn't fun for you, but rest assured, I'm not having any fun either."

Alas, we live in an imperfect world. Inflexible-explosive children are rarely able to describe their difficulties with this kind of clarity. Nonetheless, this ideal response provides a good opportunity to take a closer look at where I'm heading in this book, for it describes the downward spiral that often typifies inflexible-explosive episodes. Indeed, the child's response suggests that there is a distinct progression to many such episodes. Here's what I mean:

> When you ask me to make these shifts, I start to get frustrated . . . and when I start getting frustrated, I have trouble thinking clearly, and then I get even more frustrated.

In the early phase of an inflexible-explosive episode, the child is presented with an environmental demand to shift gears and experiences the natural frustration associated with doing so. But because of his deficits in the domains of flexibility and frustration tolerance, he has trouble tolerating this frustration and responding adaptively to this demand. It is in this early phase that he exhibits the early warning signs that he's stuck. The trouble is, he's likely to tell you he's stuck in a way that is offensive. I've come to refer to this early phase of an inflexible-explosive episode as **vapor lock** (a metaphor originated by Dr. Steve Durant, a colleague of mine at Mass General).

Let's elaborate on the theme of vapor lock for a moment by thinking of your child's brain as working in the same way as a car's engine. Vapor lock in engines is caused by excessive heat. The excessive heat causes a bubble in the gas line that prevents gas from flowing to the engine and causes the engine to stall. No matter how many times the driver pumps the gas pedal or turns the ignition, the car won't start again until the engine cools down. In a similar manner, frustration often causes a breakdown in an inflexible-explosive child's capacity to think clearly, causing him to become less coherent and rational and more overwhelmed. No matter how many times adults pump the gas pedal—scream, berate, reason, insist, reward, punish, or whatever—the child probably won't start thinking clearly again until things cool down.

A child I worked with had his own term for this phenomenon: *brain lock.* He explained that he locked on to an idea and then had tremendous difficulty unlocking, regardless of how reasonable or rational the attempts of others to unlock him. A few other children have referred to this early phase as "short-circuiting" to describe the difficulty they have at unjamming their brain circuits to think clearly and rationally. One computer-savvy child told me he wished his brain had a Pentium processor, so he could think faster and more efficiently when he became frustrated. Regardless of the term you use, the important point is that during vapor lock, a child may still be capable of rational thought; therefore, it may still be possible to prevent a full-blown catastrophe.

. . . then you guys get mad . . .

Now we've moved to the second phase of an inflexible-explosive episode: the **crossroads** phase. This is the parents' last chance to respond to their child's frustration in a manner that facilitates either (1) communication and resolution or (2) further deterioration. As you've probably found, getting mad has probably been moving your child closer to "2" than to "1" because getting mad probably doesn't reduce your child's frustration or help him think more clearly. Luckily, there's a whole universe of things you can do during this phase besides getting mad. As you shall see.

> Then I start doing things I wish I didn't do and saying things I wish I didn't say. . . . Then you guys get even madder and punish me, and it gets really messy.

A deteriorating inflexible-explosive child is neither a pretty sight nor a pleasant experience. If the child deteriorates past the point of rescue, he'll become completely overwhelmed by frustration and lose his capacity for coherent, rational thought. At that point, the final phase of the episode has been reached: **meltdown**. Others have referred to this phenomenon as "disintegrative rage"; Dr. Daniel Goleman, author of *Emotional Intelligence*, referred to it as a "neural hijacking." It is during meltdowns that inflexible-explosive children are likely to lapse into their most destructive, abusive behavior. I use the term **mental debris** to describe the horrible words that may come out of a child's mouth during these incoherent moments.

Obviously, *meltdown* is not an original term; go to any playground frequented by two year olds and you'll hear their parents describing, often with good humor, their children's latest "meltdown of the week." The meltdowns of inflexible-explosive children often look very much like those of two year olds. But the parents of inflexible-explosive children do not describe meltdowns with good humor. They've been enduring them for a long time, and the meltdowns have become much more frequent, intense, and uncontrollable.

The trouble is, little or no learning occurs for a child while he is in the midst of a meltdown (as far as I can tell, few people are receptive to learning when they are in an incoherent state). In fact, attempts to continue to teach the child how to behave while he is in the midst of a meltdown—even trying to talk reasonably—have an excellent chance of increasing his frustration and making it that much more difficult for him to regain coherence. Punishing a child during a meltdown is likely to fuel his frustration even further and, as you may have found, may not decrease the odds of a meltdown the next time he's frustrated.

> After the dust settles—you know, when I start thinking clearly again—I end up being really sorry for the things I did and said.

This statement speaks to the fact that once their coherence has been restored, these children often express deep remorse for what they've said or done, although many have difficulty recalling events that occurred during the meltdown or what they were so upset about in the first place.

OK, now you have a jump-start on a new way of interpreting and describing your child's behavior and an appreciation for the general sequence of events. But this sequence can be useful in another way: It can help you think about *when* you should be intervening, and once that's been established, *how* you should do it. You'll get more details on the when and how as you continue reading. Briefly, though, what I've found is that many parents of inflexible-explosive children put most of their energy into intervening during and after the meltdown phase. I call this reactive intervention, or intervening on the back end, and, in the case of inflexible-explosive children, I have little faith that intervening at these points will be productive. The emphasis in this book is on helping you respond to your child before he's at his worst—in other words, proactively, on the front end. Thus, an important focus of this book is to help you learn how to anticipate and recognize situations that routinely lead to vapor lock—so you can respond before vapor lock sets in—and to help you become more aware of the things you can do during the vapor lock and crossroads phases to prevent meltdowns.

❄

You've just been given a lot of new ideas to digest. Perhaps the following summary of this chapter will be helpful:

- Flexibility and frustration tolerance are critical developmental skills that some children fail to develop as they move beyond the Terrible Twos. Inadequate development of these skills can contribute to a variety of behaviors—sudden outbursts, prolonged tantrums, and physical and verbal aggression, often in response to even the most benign of circumstances—that frequently have a traumatic, adverse impact on these children's interactions and relationships with parents, teachers, siblings, and peers.

- How you interpret your child's inflexible-explosive behavior and the language you use to describe it will directly influence the strategies you use to help your child change this behavior.

• Putting your old interpretation on the shelf will also mean putting your old parenting practices on the shelf. In other words, helping your child be more flexible usually means that you will have to be more flexible first. This isn't as unfair or unreasonable a statement as it may sound, for you've already experienced the futility of responding inflexibly to a child who is, by nature, inflexible. I've often used a simple equation to capture this phenomenon:

inflexibility + inflexibility = meltdown

• One of the most important things you can do to help your child explode less often is to approach his difficulties proactively, rather than reactively—as you shall see.

3

Pathways to Inflexibility-Explosiveness

The previous chapter provided a brief overview of the various neurobiochemical irregularities that may contribute to inflexibility and explosiveness. The disorders and learning inefficiencies that these irregularities give rise to can compromise your child's capacity to think through problems; deal effectively with curves in the road; and handle frustration in a flexible, adaptive manner. In other words, a variety of pathways—including difficult temperament, ADHD-executive function deficits, social skills deficits, language-processing issues, nonverbal learning disabilities, and sensory integration dysfunction—can set the stage for a chronic pattern of vapor lock and meltdowns. It's probably more important for you to understand how each of these pathways can lead to inflexibility and explosiveness than which precise neurotransmitters and regions of the brain may be involved. So let's take a closer look at each.

Difficult Temperament

To many of us in the mental health professions, the connection between difficult temperament and problem behavior is crystal clear; just don't ask us for a simple description of what we mean by "temperament." Fortunately, Dr. Stanley Turecki, a psychiatrist in New York, has written an excellent book on children with difficult temperaments entitled *The Difficult Child*. Turecki defines temperament as *the natural, inborn style of behavior of each individual* and adds that this style of behavior is *innate and . . . not produced by the environment*. On the basis of the work of Drs. Alexander Thomas, Herbert Birch, and Stella Chess, Turecki delineated nine characteristics of temperamentally difficult children: high activity level, distractibility, high intensity, withdrawal or poor reaction to new or unfamiliar things, poor adaptability (reacting badly to changes in routine), negative persistence (strong willed, whiny, rigid), low sensory threshold (for example, clothes that don't "feel" right), and negative mood (cranky, irritable).

Does this sound like your child? Here's something you probably already know: These temperamental features are often evident, in one form or another, in infancy (although I've come across a fair number of children who were described as relatively "easy" infants but became much more difficult somewhere between one and four years old). Temperamentally difficult infants react badly to changes in routine; are restless and squirmy; protest when first introduced to new foods, places, or people; may be fussy or colicky; are poorly regulated with regard to feeding and sleeping; and are easily bothered by and overreactive to noises, lights, and other stimuli. I've yet to hear a temperamentally difficult infant described as "manipulative," "coercive," "bratty," or "controlling." Rather, it is generally assumed that temperamentally difficult infants are simply built that way. We must therefore also ponder why the explanations become a lot less compassionate as children get older. I often tell parents that their inflexible-explosive child is simply a slightly older version of the temperamentally difficult infant or toddler they were blessed with a few years back.

How are difficult temperament and inflexibility-explosiveness connected? Here's the basic recipe: Mix high intensity with a poor reaction to new or unfamiliar things, add poor adaptability and negative persistence, and stir in low sensory threshold and negative mood. Season with hyperactivity (optional). Heat (briefly). You've just cooked up an inflexible-explosive child. Inflexibility-explosiveness may best be viewed as the most toxic manifestation of a difficult temperament.

In keeping with the credo that few things in psychology are always true, not all inflexible-explosive children have or had difficult temperaments. Nonetheless, you'll see a lot of overlap between the characteristics of temperamentally difficult children and the behaviors, described next, that define many of the diagnostic labels that end up on their insurance forms.

ADHD and Executive Function Deficits

A lot of temperamentally difficult children, including many of the inflexible-explosive children I work with, are diagnosed with ADHD at some point along the way. ADHD is the diagnostic label used to describe children who exhibit developmentally extreme levels of inattention or hyperactivity-impulsivity, or both. ADHD has received more research attention than any other childhood psychiatric disorder. But some of the most exciting research on ADHD has occurred over the past fifteen years. This research has shown that children with ADHD often have deficits in a crucial set of thinking skills that are commonly referred to as "executive functions." There's no universal agreement on the specific skills that constitute executive functions nor a complete consensus on the regions of the brain that govern these skills (although much research has zeroed in on the frontal, prefrontal, and frontally interconnected subcortical regions of the brain, fueled by the neurotransmitter dopamine). However, executive function skills can help us understand not only the behaviors that are included in the diagnosis of ADHD—such as difficulty sustaining attention and effort, lack of attention to details, seeming not to listen, difficulty following through on tasks or instructions, disorganization, distractibility, motoric restlessness, talking excessively, difficulty awaiting one's turn, and interrupting or intruding on others—but the poor tolerance for frustration, inflexibility, temper outbursts, and unstable mood that are commonly seen in inflexible-explosive children.

Specifically, executive functions include cognitive skills such as a shifting cognitive set (the ability to shift efficiently from one mind-set to another), organization and planning (including anticipation of problems, formulation of goals in response to problems, and selection, monitoring, and adjustment of strategies in response to problems), working memory (information stored and accessible in short-term memory), and separation of affect (the ability to separate your emotions from your thoughts). What I like about these terms is that they help us understand what's going on (or, perhaps more accurately, what's *not* going on) inside the heads of inflexible-explosive children. Let's take a closer look at each.

Moving from one environment (such as recess) to a different environment (such as a reading class) requires a shift from one mind-set ("In recess it's OK to run around and make noise and socialize") to another ("In reading, we sit at our desks and read quietly and independently"). If a child has difficulty shifting cognitive set, there's a good chance he'll still be thinking and acting like he's in recess long after reading class has started. This circumstance helps explain why children with ADHD have such trouble making transitions from the rules and expectations of one activity to the rules and expectations of another.

"I may want *to sit. Just give me a chance to process this."*

But inefficient shifting of cognitive set may also explain why a child may become so frustrated when, for example, his parents ask him to stop watching television and come in for dinner. If a child is an inefficient "shifter" and if other factors—like the parents insisting that he comply with their requests or shift gears quickly—compound the child's mounting frustration or compromise his capacity for clear thought, even seemingly simple requests may set the stage for serious explosions. Such children are not intentionally trying to be noncompliant; rather, they have trouble flexibly and efficiently shifting from one mind-set to another.

How do we know a child is having difficulty shifting cognitive set? He tells us! That's the good part. The unfortunate part is that he may not articulate this difficulty in an appropriate way. Let's listen in:

Parent: Please turn off the TV and come in for dinner.

Casey [*early vapor lock*]: No! My show's not over yet!

Parent: Turn off the TV and come in for dinner now!

Casey [*advanced vapor lock*]: I can't come in!

Parent: What do you mean, you "can't come in"?
Dinner's getting cold . . . get in here!

Casey [*crossroads*]: Shut up! I'm not listening!

Organization and planning skills are also critical to the process of thinking through one's options for working toward a goal or completing a task. Children with ADHD often have notoriously poor planning and organizational skills. This difficulty probably explains why many of them have so much trouble doing things like bringing the appropriate homework materials home from school, completing homework, and getting ready for school in the morning—tasks that involve the organization and planning of multiple sequenced smaller tasks.

However, poor organizational and planning skills may also explain the difficulty many children with ADHD have in responding effectively to life's daily barrage of problems and frustrations. To begin with, such children are often taken by surprise at problems because they're not good at anticipating them in advance, even when certain problems seem to occur over and over. Surprised or not, to respond to a frustrating situation, one must efficiently organize a coherent plan of action—that is, analyze the problem; formulate a goal; and select, monitor, and adjust one's strategy. This plan of action would presumably incorporate prior experience ("How have I successfully handled this problem before?" "What previous actions have I taken in response to this problem that have *not* been as successful?") and take into account various situational factors ("My mom seems really angry this time!" "The bus will be here in three minutes; I'd better hustle!"). Thus, if a child wants to watch television a few more minutes before coming in for dinner, he must efficiently—within a few seconds—organize and plan a coherent response to a parent who is expecting rapid compliance. If the child's organization and planning skills are compromised, his response to this situation and many others like it could be disjointed and disorganized and would likely be accompanied by a debilitating level of frustration. Such children are not intentionally trying to be noncompliant; rather, they have trouble flexibly and efficiently organizing coherent plans of action in response to problems or frustrations.

Parent: My child does just fine unless something doesn't go his way.

Me: Precisely.

An executive skill that operates in tandem with organization and planning is working memory. Working memory is a person's capacity to "hold that thought" until his brain has had a chance to think something through. Poor working memory may help explain why many children with ADHD interrupt and intrude on others and blurt out answers in class without being called upon. As Jack, a child with ADHD, told me, "If I have to wait until I'm called on, I'll forget what I wanted to say . . . and if I try to remember what I want to say, I won't be able to follow the discussion that's going on, so when I'm finally called on, my answer seems off-base."

Poor working memory may also explain why many children with ADHD have difficulty responding adaptively to frustrating problems. If you're unable to maintain a mental representation of the problem ("My mom wants me to come in for dinner, but I want to watch TV for a few more minutes") in your thoughts while you're trying to think through potential solutions, the "thinking-through" process is likely to stall, and you're likely to respond with your first impulse at resolving a problem. And, unfortunately, your first impulse usually is not the best way to respond. Few problems are simple enough to have straightforward solutions; rather, most problems we humans face are complex and have similarly complex solutions. In other words, most problems aren't successfully resolved with one quick thought. Thus, resolving problems usually requires us to maintain the idea of the problem in our minds while we also hold in our minds the different variants of solutions we've generated until we decide on a plan of action that reflects the evolution of our thinking about the problem. And since many initial action plans don't work out perfectly, we usually have to shift cognitive set yet again and continue thinking things through several times before the problem is satisfactorily resolved. Children with poor working memory are not intentionally trying to be noncompliant; rather, they have trouble efficiently and flexibly sorting through different solutions to frustrating problems. A typical scenario may sound something like this:

Parent: I'm running a little behind today. Finish your breakfast, put your dishes in the sink, and get ready for school.

Child [*early vapor lock*]: But I'm not through eating yet.

Parent: Why don't you grab an apple or something. Come on, hurry! I have to drop some things at the post office on the way there.

Child [*advanced vapor lock*]: I can't do that!

Parent: You can't do what? Why do you always do this when I'm in a hurry? Just this once, could you please do what I say without arguing?

Child [*crossroads*]: I don't know what to do!

Parent: I just told you what to do! Don't push me today!

Child: [*meltdown*]

Thinking clearly is a lot easier if a child has the capacity to separate or detach himself from the emotions elicited by frustration, an executive skill sometimes referred to as "separation of affect." This skill permits people to put their emotions on the shelf so as to think through solutions to problems more objectively, rationally, and logically. Children whose skills in this domain are lacking or inefficient become overwhelmed with frustration in response to even simple problems and respond to frustrating situations with less thought and more emotion. They may actually feel themselves "heating up" but often aren't able to stem the emotional tide until later, when the emotions have subsided and rational thought has kicked back in. They may even have the knowledge to deal successfully with problems (under calmer circumstances, they can actually demonstrate such knowledge), but their frustration prevents them from remembering the information when they need it the most. Such children are not intentionally trying to be noncompliant; rather, they become overwhelmed with the emotions associated with frustration. You know what this pattern is like:

Parent: It's time to stop playing Nintendo and get ready for bed.

Child [*vapor lock*]: Damn it! I'm right in the middle of an important game!

Parent: You're always right in the middle of an important game. Get to bed! Now!

Child [*crossroads*]: Shit! You made me mess up my game!

Parent: I messed up your game? Get your butt in gear before I mess up something else!

Child: [*meltdown*]

Executive function deficits may also help explain why many children with ADHD think in a concrete, black-and-white manner. You see, while some situations require minimal cognitive processing ("Johnny, sharpen your pencil and sit back down at your desk"), others require more complex processing ("Johnny, I want you, Francis, Lisa, and Bob to work together and come up with a presentation on the early explorers of Florida"). The former situation usually requires only a rote response, which is easier for an inefficient thinker. But inefficiency of thought may cause such children to respond to the latter situation in a concrete manner as well ("I can't do this! They're not doing the presentation the right way!"), demonstrating an incapacity to deal adaptively with circumstances that involve greater complexity and require greater cognitive efficiency.

Executive function deficits can also explain why children with ADHD may demonstrate so little insight about themselves and equally minimal awareness of the impact of their behavior on other people. Indeed, children with executive function deficits often come off as demanding, self-centered, and lacking in empathy and social tact. Self-awareness and empathy require a fairly continuous process of reflecting on and rethinking of experiences and perspective taking, and this process may be short-circuited by deficits in executive skills. As you'll read in later chapters, helping inflexible-explosive children achieve a basic level of insight into their difficulties can help them respond to frustration more adaptively. Of course, helping *you* achieve greater insight into these difficulties is even more important right now.

Social Skills Deficits

I can think of few human activities that require more flexibility, complex thinking, and rapid processing than social interactions. Thus, it should come as no surprise that children whose thinking is rigid and inefficient find their social interactions to be frustrating and this frustration may fuel many of their explosions. In collaboration with several of his colleagues, Dr. Ken Dodge, a researcher at Vanderbilt University, has delineated a set of specific thinking skills—known as "social information processing skills"—that are thought to be involved in even the most trivial social interactions. A brief review of these thinking skills will help you understand just how labor-intensive social interactions can be, especially for children who aren't good at them.

Let's say a boy is standing in a hallway at school and a peer comes up and, with a big smile on his face, whacks the boy hard on the back and says, "Hi!" The boy who was whacked on the back now has a split-second to attend to and try to pick up on the important qualities of the social cues of this situation ("Who just whacked me on the back? Aside from the person's smile, is there anything about his posture or facial expression or this context that tells me whether this was a friendly or unfriendly smile and whack?"). At the same time he must connect those cues with his previous expriences ("When have other people, and this person in particular, whacked me on the back and smiled at me before?") so as to interpret the cue ("Was this an overexuberant greeting or an aggressive act? How do I feel about it?"). Then he has to think about what he wants to have happen next ("That was a mean thing to do. . . . I'd like to avoid getting into a fight with this person" or "That was a nice greeting. . . . I'd like to play a game with him"). Then, on the basis of his interpretation of the cue and the outcome he desires, the boy must begin to think about how to respond, either by remembering his experiences in similar situations or by thinking of new responses. Then, he must evaluate the different possible responses, consider the likely outcomes of each ("If I smile back, he'll probably ask me to play a game with him"), choose a response, enact it, monitor the course of events throughout, and adjust the response accordingly.

It sounds like a lot of thinking for one event, yes? The key point is that this process is nonstop and requires a lot of efficiency. It's barely noticeable to people who do it without even thinking about it, but it's very frustrating when it doesn't happen automatically. As children get older, social interactions become increasingly complex and require even greater efficiency of thought.

Many inflexible-explosive children are unskilled at recognizing the impact of their behavior on others; may apply the same rigid, rote interpretation (for example, "He doesn't like me" or "That's not fair") to diverse, complex social information; and may have a limited repertoire of responses and end up applying the same response (hitting, screaming, crying, exploding) to a wide range of social situations. If a child isn't "built" to reflect on the accuracy of his interpretations, the effectiveness of a given response, or the manner in which his behavior affects others—a circumstance that Daniel Goleman referred to as "emotional illiteracy"—he is likely to find social interactions extremely frustrating. This can, at the least, contribute to the child's general level of frustration; at worst, it may lead to a chronic pattern of explosions. There is an excellent book by Drs. Stephen Nowicki and Marshall Duke called *Helping the Child Who Doesn't Fit In* that describes these social difficulties—and how to help—more fully.

Sometimes a child's social functioning is so severely impaired that the diagnosis of pervasive developmental disorder (PDD) may be considered. PDD is the category under which disorders such as autism and Asperger's disorder are subsumed. These disorders are characterized by both marked delays in social interaction and communication skills and greatly restricted or repetitive interests, activities, and behaviors. Depending on the specific form of PDD, these features may occur with or without impaired functioning in other cognitive domains. As you may imagine, the constellation of difficulties typifying PDD can also severely compromise a child's ability to deal adaptively with frustration.

Language Processing

It may have occurred to you as you read the foregoing that a lot of the thinking and communicating we do involves language. Many prominent theorists—such as Lev Vygotsky, Jean Piaget, and Aleksandr Romanovich Luria—have underscored the importance of language skills in setting the stage for many forms of thinking, including reflecting, self-regulating, goal setting, problem solving, and managing emotions. Thus, as you may imagine, children who are less efficient at understanding language, categorizing and storing current and previous experiences (in language), thinking things through (in language), or expressing themselves (via language) are at risk of being less effective in dealing adaptively with frustration.

Let's take these basic language skills one at a time. Children who have difficulty understanding the complex array of verbal events going on around them often become confused and frustrated by these events. In other words, understanding language is often the first necessary step in responding to people around you and the world in general. It's hard to formulate an adaptive response to a given situation when you don't fully comprehend things in the first place.

Some children understand language just fine but have trouble categorizing and labeling their emotions. This is another step needed for responding to the world in an adaptive manner. If a child isn't skilled at categorizing or labeling previous experiences, he may not know how he's feeling at any given moment and may therefore have difficulty remembering how he previously responded when he felt the same way. Can you imagine *feeling* all the sensations associated with frustration—hot-faced, agitated, tense, explosive, and so on—without being able to label and categorize what you're feeling? You'd probably become even more frustrated, and your ultimate response would likely be quite counterproductive. I've worked with many children who lacked even a rudimentary vocabulary for describing their emotions. Helen, age seven—about whom you'll read more in Chapter 4—was one such child:

Me: Helen, what types of things make you happy?

Helen [*long pause*]: I don't know.

Me: Well, then, what types of things make you sad?

Helen: [*shrug*]

Me: Before you came in today, your mom told me about how she got mad at you for leaving your clothes on the floor in your room. How did her being mad at you make you feel?

Helen: I don't know.

Me: Hmm. From what I hear, you became very upset and ran out of the house when she got mad at you. You don't know what you were feeling?

Helen [*long pause*]: Upset?

Me: When you say you were upset, what do you mean?

Helen: I don't know.

Because children like Helen have a limited repertoire of "feeling" words, they may use other words in the midst of frustration instead, such as "I hate you," "Shut up," "Leave me alone," or worse. In Chapter 9, I'll talk about things that can be done to expand the "feeling" vocabularies of such children.

Some children do just fine at understanding language and labeling their feelings but have trouble thinking things through to arrive at an adaptive response. These children may be good at telling you, "I'm very *mad* right now," but may become confused and frustrated when they try to think about what to do next. All that some of them can "think" of (if you want to call it thinking) to do next when they're frustrated is to hit someone, break something, or call someone a name. What they probably need most from adults is some help in expanding their "what to do next when you're mad" repertoire.

Me: George, I understand you got pretty frustrated at soccer the other day.

George: Yep.

Me: What happened?

George: The coach took me out of the game, and I didn't want to come out.

Me: I understand you told him you were very mad.

George: Yep.

Me: I think it's probably good that you told him. What did you do next?

George: He wouldn't put me back in, so I kicked him.

Me: You kicked the coach?

George: Yep.

Me: What happened next?

George: He kicked me off the team.

Me: I'm sorry to hear that.

George: I didn't even kick him that hard.

Me: I guess it wasn't important how hard you kicked him. I'm just wondering if you can think of something else you could have done when you were mad besides kick the coach.

George: Like what?

Me: I don't know; I wasn't there. Can you think of any times you've been mad and did something that made things turn out better than getting kicked off the soccer team?

George: Nope.

Of course, life isn't always so simple as "I'm mad" anyway. If I child says, "I'm mad," the world—his parents, teachers, siblings, peers, soccer coaches—is bound to respond in some way. Then the child has some more thinking, feeling, and expressing to do. The trouble is, "I'm mad" is about all some children can muster in the "expressing your feelings" department. So if the world asks for clarification or a more sophisticated articulation of "I'm mad" or demands additional thinking through, these children may become confused, disorganized, overwhelmed, and—you guessed it—frustrated.

Me: Thinking back, has George ever responded to frustration in an adaptive way?

Father: Now that you mention it, no.

Mother: But it was never this bad.

Me: How do you mean?

Mother: When he was smaller, he didn't swear at us like he does now.

Me: What did he do instead?

Mother: Well, instead of screaming things like "Fuck you!" he'd scream things like "I hate you!" or "I want to kill you!"

Me: But, when he's been frustrated, he's never said things like, "I don't know what to do," "That's bothering me," or "I need help"?

Father: Absolutely not.

Mother: No way. So is there any hope?

Me: Sure there's hope. Just don't expect him to develop a vocabulary he's never had overnight.

Mood

After a long day at work, I'm often hungry, tired, emotionally spent, and irritable. Any problems or frustrations— even those that require minimal thought—seem like major obstacles. My sense is that most of us have our irritable, agitated, tired moods; if we're lucky, these moods are relatively short-lived, and people around us know what to do to help restore our "normal selves." But there are some people—including children—who are *constantly* in an irritable, agitated, cranky, fatigued state of mind. This chronic mood state can severely compromise a person's capacity to be flexible and deal adaptively with frustration, and children who are so afflicted may react to even minor problems and frustrations as if they were major obstacles.

Are these children depressed? Some mental health professionals reserve the term *depression* for children who are routinely blue, morose, sad, and hopeless, which actually seems *not* to be the case with many of the irritable inflexible-explosive children I know. Some inflexible-explosive children who exhibit the constellation of extreme irritability, agitation, and mood instability are instead diagnosed as having bipolar disorder or manic-depression. This is still a controversial diagnosis in children; it has been suggested that what some researchers and clinicians call bipolar disorder is similar to what others call "difficult temperament" and still others call "really bad ADHD" or "ADHD plus depression." Regardless of the label one chooses, in my experience—and the research of my colleagues at Mass General confirms this impression—children who meet criteria for bipolar disorder exhibit more severe behavior, are more impaired psychosocially, and have poorer long-term outcomes than do those with ADHD alone.

When children don't seem to meet the official criteria for either depression or bipolar disorder but still exhibit some unmistakable components of one or the other, my tendency is to suggest that there is a clear "mood component" contributing to their difficulties. What's crystal clear is that the inflexibility-explosiveness of these children is being fueled by a chronic state of irritability, agitation, volatility, and mood instability that makes it hard for them to respond to life's routine frustrations in an adaptive, rational manner.

Mother: Mickey, why so grumpy? It's a beautiful day outside. Why are you indoors?

Mickey [*slumped in a chair, agitated*]: It's windy.

Mother: It's windy?

Mickey [*more agitated*]: It's windy. I hate wind.

Mother: Mickey, you could be out playing basketball, swimming . . . you're this upset over a little wind?

Mickey [*very agitated*]: It's too windy, dammit! Leave me alone!

Anxiety

I used to be flight anxious. That's right, scared of flying. I'm still not the world's greatest flyer, but I'm nowhere near as bad as I used to be. No, I wasn't intentionally being anxious (sweaty palms, racing heart, catastrophic thoughts) so flight attendants would pay attention to me. I was truly unnerved to find myself five miles above the earth going 500 miles per hour in an aluminum contraption filled with gasoline, with my life in the hands of people (the pilots and air traffic controllers) I'd never met. To control this anxiety, I engaged in a few important rituals to ensure the safe progress of my flight: I had to sit in a window seat, so I could scan the skies for oncoming aircraft, and had to review the emergency instruction card before the plane took off. I knew these rituals worked because all the flights I'd been on had delivered me safely to my destination.

Did these rituals cause me to behave oddly at times? You bet. On one flight, the plane was cruising along at 33,000 feet or so and I was, as usual, vigilantly scanning the horizon for threatening aircraft. Then the unthinkable happened: I spotted an aircraft far off in the horizon ascending in the direction of my airplane. By my expert calculation, we had about five minutes before the paths of the two planes crossed and my life would come to an abrupt, fiery end. So I did what any very anxious, somewhat irrational, human being would do: I rang for the flight attendant. There was no time to spare. "Do you see that airplane down there?" I sputtered, pointing toward the speck miles off in the distance. She peered out the window. "Do you think the captain knows it's there?" I demanded. The flight attendant tried to hide her amusement and said, "I'll be sure to let him know." I was greatly relieved, albeit certain that my heroism was not fully appreciated by either the flight attendant or the passengers seated near me (who were now scanning the aircraft for empty seats to move to). As I was leaving the airplane at the end of the flight, the flight attendant and pilot were waiting at the door and smiled as I approached.

The flight attendant tugged on the pilot's sleeve and introduced me: "Captain, this is the gentleman who was helping you fly the plane."

The point is that severe anxiety has the potential to make rational, coherent thought much more difficult. And as fate would have it, it's when we're severely anxious about something that clear thinking is most crucial. This combination of anxiety and irrationality causes some children (the lucky ones) to cry. But a substantial number of them (the unlucky ones) explode. We adults tend to take things far less personally and respond far more empathically to children who cry when they're frustrated instead of explode, even though the two behaviors often flow from the same source. Also, I've become convinced that some of the obsessive-compulsive children I've seen began ritualizing because in the absence of rational thought, the rituals were the only things they could come up with to reduce their anxiety, albeit only temporarily.

I'm proud to note that I no longer feel compelled to sit next to the window or review the emergency manual and have survived flights on which I did neither. How did I get over my flight anxiety? Experience. And by thinking. An Air Florida pilot got the process going. As I was boarding this Air Florida flight, the captain was greeting passengers at the door of the aircraft. I seized the opportunity. "You're going to fly the plane safely aren't you?" I sputtered. The pilot's response was more helpful than he knew: "What, you think *I* want to die, buddy?"

That the pilot wasn't particularly interested in dying was an important revelation, and it got me thinking. About the thousands of planes in the air across the world at any given time and the slim odds of something disastrous happening to the plane that I was on. About the millions of flights that arrive at their destinations uneventfully each year. About the hundreds of flights I have been on that arrived safely. About the thousands of flights my brother and brother-in-law have flown without incident. About how calm the flight attendants look. About how many of my fellow passengers are fast asleep. Even when there's turbulence.

My flight anxiousness never caused me to become explosive. But some children do explode when they get anxious. They stop thinking clearly, become overwhelmed with anxiety, and veer off right into an explosion.

Nonverbal Learning Disability

Although any learning disability has the potential to make life a lot more frustrating for a child, one learning disability may actually set the stage for inflexible-explosive behavior. Nonverbal learning disability has drawn increased attention from researchers, and refers to a syndrome in which children have poorer nonverbal than verbal skills (often seen in the form of poorer Performance versus Verbal scale scores on the Wechsler Intelligence Scale for Children); poor mathematical skills; difficulty comprehending reading material despite good skills at reading single words; poor nonverbal memory and visual perception; significant difficulty on tasks or in situations requiring problem solving, flexibility, and adaptability (especially in response to abstract or unfamiliar stimuli); and difficulties in social perception, social judgment, and social interaction skills. Some children do not manifest the full syndrome but still exhibit several of these difficulties.

Children with this syndrome tend to be strong at learning

rote material and often approach the world in a similarly rigid, concrete manner. Unfortunately, the world—and social interactions in particular—require a lot more problem solving, flexibility, and adaptability than rote memorization and the application of concrete skills. Thus, children with nonverbal learning problems may experience enormous frustration as they struggle to apply concrete rules to a world in which few such rules apply. Their inflexibility-explosiveness makes perfect sense, given the general way they think and approach the world.

Child [*in a car*]: Dad, this isn't the way we usually go home. [*early vapor lock*]

Father [*driving*]: I thought we'd go a different way this time, just for a change of pace.

Child [*advanced vapor lock*]: But this isn't the right way!

Father: I know this isn't the way we usually go, but it may even be faster.

Child [*crossroads*]: We can't go this way! It's not the same! I don't know this way!

Father: Look, it's not that big a deal to go a different way every once in a while.

Child: [*meltdown*]

Sensory Integration Dysfunction

One final area worthy of mention as a potential pathway to inflexibility-explosiveness is sensory integration dysfunction. In addition to helping us humans sort through our thoughts, the central nervous system is also important for helping us screen, sort through, and respond to—in other words, integrate—the barrage of sensory information from the external environment. As children mature, their capacity to sort through incoming sensations and ability to direct their own motor movements develop rapidly. These skills do not develop as readily in some children, however. Occupational therapists—the professionals who generally assess and diagnose sensory integration disorders—typically focus on the different ways in which a child may be having difficulty integrating various sensory systems, such as touch, movement and coordination, body position and awareness, sight, and sound. Children who are underreactive to various sensations may fluctuate between under- and overresponsiveness, whereas children who are overly sensitive to these sensations may avoid certain clothes and textures, overreact to loud noises, and react poorly to seemingly routine movements. Sensory integration difficulties may have adverse effects on motor planning (such as using scissors, getting on a bicycle, and manipulating eating utensils), academic skills (like handwriting), and self-care skills (including tying shoes and buttoning and zipping clothes), and such children may have strong preferences for clothing (such as being sensitive to collars hitting the neck, sleeves hitting the wrists, and seams and tags in clothing) and food (for example, being oversensitive to temperature, texture, and certain flavors). Although the overlap between sensory integration dysfunction and executive functions is not well studied, a variety of behaviors typically associated with deficits in executive functions—disorganization, distractibility, poor planning skills, failing to anticipate the results of one's actions, difficulty adjusting to new situations or following directions, frustration, and aggression—have also been attributed to sensory integration dysfunction. Regardless of the label one applies, many of the inflexible-explosive children with whom I've worked have sensory integration problems, and these difficulties caused them significant frustration and clearly compromised their ability to be flexible and to tolerate frustration. A new book called *The Out of Sync Child* by Carol Stock Kranowitz describes sensory integration dysfunction in greater detail, along with how children so diagnosed can be helped.

Other Factors Contributing to Cumulative Frustration

We all have a "frustratometer"—that unofficial meter inside our heads that gauges our present and cumulative levels of frustration. Although the pathways just described can best be conceptualized as contributing to a limited-capacity frustratometer, other issues may be important because they add to a child's cumulative level of frustration. In other words, while a difficult temperament; irritable or unstable mood; anxiety; and deficits in executive functions, social skills, language processing, nonverbal skills, and sensory integration can make it difficult for a child to be flexible and to tolerate frustration, a variety of other problems may make things even worse by pushing a child that much closer to the edge.

Learning disabilities—in reading, math, and writing skills, for example—can add substantially to a child's cumulative level of frustration. A child who has a lot of vocal tics may be frustrated and embarrassed by them and may expend a substantial amount of energy trying to inhibit them. An inflexible-explosive child of divorced parents may have difficulty adapting to the disparate demands imposed by different home environments. One wonderful boy with whom I was working was very short; although he did a good job of dealing with the teasing and his own self-doubts most of the time, his short stature continually contributed to his cumulative level of frustration. Difficult interactions with a teacher, parent, stepparent, or coach can add to a child's frustration; so can difficult interactions with peers or siblings or poor athletic skills. Although many of these frustrations may be considered normal for most children, a limited-capacity frustratometer can make them feel like an insurmountable burden.

It should be clear that multiple factors may compromise a child's capacity for flexibility and frustration tolerance. Perhaps the most important thing the pathways described in this chapter can help you understand is that flexibility and frustration tolerance are not skills that come naturally to all children. We tend to think that all children are created equal in these capacities, and this tendency causes many adults to believe that inflexible-explosive children must not *want* to be compliant and handle frustration in an adaptable way. As you now know, in many cases, this simply isn't true.

It should be clear that multiple factors may compromise a child's capacity for flexibility and frustration tolerance. Perhaps the most important thing the pathways described in this chapter can help you understand is that flexibility and frustration tolerance are not skills that come naturally to all children.

It's important to get a good handle on the precise pathways that appear to be contributing to your child's inflexibility-explosiveness. These pathways will have implications for the interventions you apply to try to help your child be more flexible and handle frustration more adaptively. Indeed, it's hard to imagine trying to help your child without the fullest possible understanding of his difficulties. How do you gain such an understanding? With professional assistance, although your impressions and knowledge of your child will be a critical part of the process. Given the wide range of factors that may be fueling your child's difficulties—many of which require a specialized assessment—my general recommendation is that you have your child comprehensively assessed by a competent child psychologist or neuropsychologist. You don't want to leave any stone unturned. Here are the components that should be integral parts of such an assessment (the sequence is less important, as long as all the components are included).

Information about your child's **developmental, school, and treatment history** is crucial and is typically obtained through interviews with the child's parents. The assessor will want to know about your child's prenatal history, early development and attachment, whether your child's difficulties with inflexibility and low frustration tolerance are longstanding; whether your child has a history of difficult temperament and learning problems; whether there is a history of abuse or trauma; and the types of treatment, if any, your child has received for these difficulties. Information you can provide about the **psychiatric history** of family members (parents, grandparents, siblings, and so forth) can provide what may best be described as a historical backdrop to your child's difficulties, along with diagnostic clues. Although arriving at a diagnosis should not be the primary goal of the evaluation, the assessor should also obtain information about your child's past and current behavior during a **diagnostic interview,** covering all significant childhood psychiatric diagnostic categories including ADHD, ODD, OCD, PDD, mood and other anxiety disorders, and social functioning. The **situational aspects** of your child's inflexibility-explosiveness—with whom, in what settings, under what circumstances, and at what times of the day meltdowns routinely occur—should also be assessed. The assessor should ask you to complete **behavior checklists** to determine the degree to which your child's behavior differs from that of other children of the same age and gender. Finally, your understanding of and reactions to your child's inflexible-explosive behavior should be informally assessed.

The assessment process should also include an interview with your child, to assess his understanding of and views about his difficulties. Your child's **emotional functioning** should be formally or informally assessed as well. The assessor will also want to get a good sense of **family dynamics and interactions**. To do so, the assessor may want to include siblings, grandparents, and other important family members in the process.

Information about your child's **functioning at school** and in other relevant settings is also essential. Toward this end, the assessor should ask teachers and other adults who have extensive contact with your child to complete **behavior checklists**, so as to obtain different points of view regarding his behavior and to assess further the situations in which inflexible-explosive behavior does and does not occur. If such behavior occurs frequently at school, teachers and other school staff should be interviewed in the same manner as parents; sometimes they can be interviewed by phone, although this is not the preferred mode of communication.

A formal psychoeducational evaluation should be performed to obtain crucial information regarding other pathways. Such an evaluation should include an assessment of your child's **general cognitive functioning**, to determine unevenness in verbal versus nonverbal skills and to gain a sense of your child's general abilities; **achievement skills**, to determine the presence of additional learning disabilities in academic skill areas, such as reading, mathematics, and writing; and **executive function, language processing, motor,** and **sensory integration skills**.

Why are you putting your child and yourself through all this? Because it's hard to know how to begin to help your child until everyone—parents, other family members, teachers, soccer coaches, and mental health professionals—has a much clearer understanding of what's fueling his inflexibility and explosiveness. An evaluation aimed solely at determining whether your child qualifies for special education services is unlikely to produce this comprehensive understanding.

It's time for another car analogy. Let's say the engine in my car began making a lot of strange noises, and I decided to take it to a mechanic. The mechanic wouldn't have a precise idea of what parts of my engine needed fixing until he fully assessed the range of problems that might cause it to make strange noises. If my mechanic *automatically* assumed that strange engine noises were *always* accounted for by, say, a bad belt, it's likely I'd leave his shop with a brand-new belt. But, given the range of alternative explanations, there's a good chance my car would still be making strange noises. In other words, the problem probably wouldn't be fixed until it was well assessed and well understood to begin with.

Don't make the mistake of thinking that having a mental health professional bestow a formal diagnosis is all you need to achieve a complete understanding of your child's difficulties. In other words, knowing that your child "has ADHD" or "is depressed" or "has PDD" doesn't tell you a whole lot about what your child needs help with, what kind of help he needs, or how to get him that help. I'm much more interested in working toward a consensus about the specific problem behaviors a child is exhibiting, describing the behaviors accurately, understanding where the behaviors are being exhibited and what factors seem to be fueling them, and—on the basis of this comprehensive understanding—bringing to bear the different treatments that seem appropriate to the child's needs.

Not all children with difficult temperaments; mood and anxiety disorders; nonverbal learning disabilities; or deficits in executive functions, social information processing, language processing, and sensory-integration ultimately become inflexible-explosive. Perhaps it is the *severity* of a child's difficulties that leads to inflexibility-explosiveness in some instances and the *combination* of difficulties that leads to it in other instances. And, in many cases, perhaps it is the manner in which the world responds to a child's inflexibility-explosiveness that makes all the difference. This response, as you may suspect, is what the rest of the book is about.

4

Inflexible-Explosive Faces

Given the multiple pathways you just read about, inflexibility and low frustration tolerance can look different in different children. Some children have such incredibly short fuses that the slightest frustration often spins them into immediate vapor lock. Others have longer fuses but still aren't adept at thinking things through; their descent into a meltdown may be more prolonged but just as nasty once they get there. Some children are pretty good at letting you know when their circuits have started jamming ("I can't talk about that right now!"); others let you know in less desirable ways ("Shut up!" "Fuck you!" "I hate you!" or "Leave me alone!"). Some children warn of an impending meltdown not with words but with a sudden onset of irritability, whining, or fatigue; others don't give much warning at all. Many are able to hold it together at school but then melt down almost instantaneously when they get home; some melt down both at home and at school. Some children throw things or hit during meltdowns, others scream and swear, and still others cry; I've known others who curled up into a ball on the floor.

So the "new language" and characteristics you've read about start to come to life, it's useful for you to have a good idea of how inflexibility-explosiveness appeared in some of the children with whom I've worked. My bet is that you'll see similarities between the children described in this chapter and the inflexible-explosive child you're trying to parent or teach. These children and their families are revisited in one way or another throughout the book.

Casey

Casey was a six-year-old boy who lived with his parents and younger sister. In our first session, his parents reported that, at home, Casey was hyperactive; had difficulty falling asleep at night; was unable to play by himself (but had equal difficulty engaging in cooperative play); was restricted and rigid in the clothes he was willing to wear and the food he was willing to eat (he often complained that certain fabrics were annoying to him and that many common foods "smelled funny"); had a lot of difficulty with transitions (getting him to come indoors after playing outside was often a major ordeal); became anxious when presented with new tasks or situations; was frequently in an irritable, agitated mood; and melted down many times each day. Most of these characteristics had been present since Casey was a toddler.

His parents noted that Casey seemed to be bright, in that he had excellent memory for factual information, and thought that his fine and gross motor skills were somewhat delayed compared to other boys his age. Casey had never undergone a complete psychoeducational evaluation, but was evaluated by an occupational therapist when he was three years old. The therapist thought that Casey had difficulties with sensory motor integration and began working with him in occupational therapy at that time. This therapy helped Casey feel a greater sense of mastery over the barrage of sensory stimuli he had to deal with, but his meltdowns continued. His parents had previously consulted a psychologist, who helped them establish a behavior management program. The parents vigilantly implemented the program but found that Casey's hyperactivity and irritability were more potent than his clear desire to obtain rewards and avoid punishments. Indeed, the program actually seemed to frustrate him further, but the psychologist encouraged the parents to stick with it, certain that Casey's performance would improve. It didn't, so the parents discontinued the program after about three months.

The parents had read a lot about ADHD and thought that this diagnosis fit Casey, but that many of his characteristics fell outside the realm of this disorder. They thought the term *control freak* fit their son better than any traditional diagnosis. They often tried to talk to Casey about his behavior, but even when he was in a good mood, his capacity for thinking about his own behavior seemed limited; after a few seconds, he would yell, "I can't talk about this right now!" and run out of the room.

"We can live with a lot of Casey's behaviors," said his mother. "But his explosions . . . and the way they disrupt our entire family . . . and our concern about what's going to happen to him if we don't help him . . . really worry us."

Casey had difficulties at school, too. His first-grade teacher reported that Casey would occasionally hit or yell at other children during less structured play or academic activities, particularly when he did not get his way. Like the parents, Casey's teacher was impressed by his factual knowledge but concerned by his difficulty integrating information— that is, combining two or more pieces of information into a coherent whole—and poor problem-solving skills. When lessons called for recall of rote information, Casey was the star of the class. When lessons required the application of this information to more abstract, complex, real-life situations, his responses were disorganized and off the mark.

When he was in a good frame of mind, Casey's communication skills seemed fine, but when he was frustrated by a particular classroom situation or task, he would yell, "I can't do this!" and would become quite agitated or start crying; sometimes he would run out of the classroom. On several occasions, he ran out of the school, which caused great concern for his safety. Sometimes he regained his composure quickly; other times it took twenty to thirty minutes for him to calm down. Afterward, Casey was either remorseful ("I'm sorry I ran out of the classroom . . . I know I shouldn't do that") or had difficulty remembering the episode altogether. The teacher reported that she could often tell from the moment Casey walked through the door in the morning that he was going to have a tough day. But she also observed that Casey was capable of falling apart even when his day seemed to be going smoothly. The teacher thought that Casey was somewhat distractible and far more hyperactive and impulsive than most boys his age, and she had hinted to the parents that Casey might benefit from medication and counseling. The parents and Casey's teacher were becoming increasingly concerned about Casey's relationships with other children; Casey seemed to lack an appreciation for the impact of his actions on others and seemed unable to use the feedback he received from others to adjust his behavior.

In my first session with Casey, he was very hyperactive and seemed unwilling or unable to talk about the important issues that he might need help with. He bounced from one toy to another in my office. When his parents were brought into the session, he settled down just long enough to hear that the reason he was in my office was because he sometimes became upset when things didn't go exactly the way he thought they would. He agreed that this was sometimes a problem. When the parents tried to get Casey to talk about this issue, he buried his face in his mother's shoulder; when the parents persisted, he warned, "I can't talk about this right now!" When they persisted further, he became red-faced and agitated and ran out of the office.

"Was that pretty typical?" I asked the parents.

"No, at home he'd have become a lot more frustrated," replied his mother. "He doesn't usually hit us—although he has hit kids at school—but he falls apart completely . . . turns red, screams or cries, yells 'I hate you!'"

"You know, in some ways his running out of the room is adaptive," I commented.

"How's that?" asked the father, a little surprised.

"Well, based on what you've told me, it seems pretty clear that he has a lot of trouble thinking and talking about his own behavior, integrating new information efficiently, and tolerating the frustration he feels when we ask him to do one of those things," I said. "While we wish he would 'use his words' to discuss things with us, his running out of the room probably keeps him from doing other things—swearing, throwing things, becoming physically threatening—that would be a lot worse."

Was Casey inflexible-explosive? It seemed so to me. Using the unofficial criteria presented in Chapter 2, he seemed to fit the bill on many counts. He clearly had a limited capacity for flexibility and adaptability and became incoherent in the midst of severe frustration; had difficulty thinking things through; had an extremely low frustration threshold and frustration tolerance; remained inflexible and easily frustrated even in the face of salient, meaningful consequences; fell apart over seemingly benign events; and had several issues about which he was especially inflexible. There appeared to be a combination of possible pathways to inflexibility-explosiveness that required further evaluation, including ADHD, mood and anxiety issues, sensory integration dysfunction, and learning problems. Thus, my short-term goal was to achieve greater clarity on these potential pathways.

My general long-term goals were (1) to provide his parents and teachers with a framework for creating environments at school and home that would set the stage for improved flexibility and frustration tolerance; (2) to help Casey feel comfortable enough, focus long enough, and stay calm enough to articulate his frustration so he could begin talking things through and be more open to adult assistance when he became frustrated; (3) to determine, over the long term, whether Casey could begin to deal with frustration more independently; and (4) to explore whether medications might be useful if the nonmedical approach proved insufficient. These goals and the strategies for achieving them are described in greater detail in later chapters.

Let's turn to another inflexible-explosive child and see what her inflexibility-explosiveness looked like.

Helen

I first met with Helen and her mother and father when she was seven years old. Helen was described as a charming, sensitive, creative, energetic, sociable girl. Her parents also described her as intense, easily angered, argumentative, resistant, and downright nasty when frustrated. They had observed that Helen seemed to have a lot of trouble making the transition from one activity to another and tended to fall apart when things didn't go exactly as she had anticipated. They reported that weekends were especially difficult; although Helen didn't love going to school, unstructured weekend time seemed to make her irritable and difficult to please. A child psychiatrist had given Helen a tentative diagnosis of ADHD, but had conceded that ADHD didn't seem to encompass the full range of behaviors Helen was exhibiting. Her second-grade teacher reported that Helen had a tendency to grumble when new lessons were introduced. Her piano teacher observed that Helen tended to become easily frustrated and often balked at trying new pieces of music. Psychoeducational testing by school personnel indicated that while Helen was of above-average intelligence, her expressive language skills were delayed.

In one of my early meetings with Helen's parents, they recounted one of her meltdowns during the previous week. "On Tuesday, Helen told me she'd like to have chili for dinner the next night," recounted her father. "So, on Wednesday afternoon, I left work a little early and made her the chili she had asked for. When she got home from swimming late Wednesday afternoon, she seemed a little tired; when I announced to her that I had made her the chili she wanted, she grumbled, 'I want macaroni and cheese.' This took me a little bit by surprise, since I know she really loves chili. It was also a little irritating, since I had put time into doing something nice for her. So I told her she would have to eat the chili. But she seemed unable to get macaroni and cheese out of her head, and I continued to insist that she eat the chili for dinner. The more I insisted, the more she fell apart. Eventually, she lost it completely. She was screaming and crying, but I was determined that she would eat the chili I had made her."

"What did you do then?" I asked.

"We confined her to her bedroom and told her she had to stay there until she was ready to eat the chili," said Helen's mother. "For the next hour she screamed and cried in her room; at one point, she was banging on her mirror and broke it. Can you imagine? All this over chili! I went up to her room a few times to see if I could calm her down, but it was impossible. Helen was totally irrational—the amazing thing is that at one point, she couldn't even remember what she was upset about."

I then asked the parents a question you'll see lots more of in later chapters. "How important was it to you that she eat the chili instead of the macaroni and cheese?"

"I couldn't have cared less," the father quickly responded, knowing where I was heading. "In retrospect, I honestly didn't care which of the two she ate." The mother confirmed this point of view.

Then came another question you're going to see a lot more of: "Do you think that your enduring this meltdown—having Helen go nuts in her room for an hour, breaking her mirror, and ruining your evening—made it any less likely that she'll melt down the next time she's frustrated over something similarly unimportant?"

"No," was the instantaneous, unanimous response.

"What was Helen like when the episode was all over?" I asked.

"Very remorseful and very loving," the mother responded. "It's hard to know whether to reciprocate her being loving or to hold a grudge for a while to cement the point that we don't like that kind of behavior."

"Well," I replied, "if you don't think that inducing and enduring meltdowns is going to help her deal better with frustration the next time, then it follows that holding a grudge probably isn't going to help either."

"Yes, but how will she learn that that kind of behavior is unacceptable?" asked the mother.

"There are a lot of things that might help, though I'm still not sure which ones," I replied. "But from what I've heard, the fact that you disapprove of that kind of behavior is pretty well cemented in her mind already . . . so I doubt that we'll be needing more cement. She also seems genuinely motivated to please you both . . . and seems as unhappy about her meltdowns as you are . . . so I'm not sure she needs additional motivation. What we have to figure out is why, under certain circumstances, a kid who knows you disapprove of her temper outbursts and is eager to please you continues to have temper outbursts that displease you."

Later that session, Helen and I played Connect Four, a game she already knew how to play. I noticed that even after several games, Helen seemed to have a lot of trouble deciding on a strategy. Each time it was her turn, she begged me to tell her what to do for her next move. Sometimes she refused to make a move without being guided on a strategy. Game after game, it seemed that although Helen plainly *understood* the rules of Connect Four, she was unable to organize and plan a coherent strategy and didn't seem able to think about potential moves for more than a moment before begging for guidance.

At the end of this session, I described to the parents my early impression of their daughter's difficulties. "Based on what you've told me and what I've observed myself, Helen seems to have a lot of trouble thinking clearly when a situation is new, unfamiliar, or uncertain or requires extensive thinking and planning. When we play Connect Four, she seems practically paralyzed unless I give her some sort of help. Have you ever noticed that she becomes very disorganized and frustrated when she's uncertain about something?"

"Absolutely," the mother replied. "When we're helping her with her homework, she needs a lot of organizing and reassurance for her to stick with tasks she's uncertain about. If we don't help her think, she starts falling apart."

"From the stories you've told me, it sounds like the same thing happens when she's uncertain in other situations," I commented. "She seems to be lacking a framework for thinking about things when she's uncertain. She's also easily overwhelmed by frustration, and once she's overwhelmed, it's practically impossible for her to think rationally until she calms down again.

"Let's start talking about what we can do to help her," I continued. "Her language-processing issues and ADHD seem to be the primary factors that are fueling her difficulties, although there may be a mood component as well. Let's talk about the language piece first. A lot of the thinking people do in their heads relies heavily on language skills. In other words, most of our thoughts are words and feelings and categories and actions that involve language. From what I've heard and seen, Helen also doesn't seem to have the words to express herself, and this is even more true when she's frustrated. Her ADHD adds disorganization and an impulsive response style to the mix. I think the testing her school did is probably fine for now, although at some point we may want to evaluate other aspects of her cognitive functioning more fully, just to make sure nothing's been overlooked.

"I think it's important for you to understand that a lot of the things Helen says or does when she's frustrated are clear evidence that she's having trouble thinking coherently and to recognize that you can play an important role in keeping her from becoming quickly overwhelmed when she's uncertain or frustrated. To assist her in this way, we'll need to work toward helping Helen come to view you both as people who can help her to think things through—rather than as adversaries—when she's starting to become upset. In other words, when she's in what I call the 'red haze of frustration,' it will be easier to restore her to rationality if she sees you as potential helpers, not as enemies. We also need to see if we can give her some sort of framework for thinking things through when she's uncertain and disorganized. That's not going to be so easy because of her difficulties with language and impulse control, but I think it can be done. If we can help her take some baby steps in this direction and then slowly add some new skills, over time I'm optimistic she'll be able to start thinking through frustrating situations on her own, rather than needing you both to do the thinking for her."

Anthony

Anthony was an eight-year-old second grader when I first met him and his parents. I remember that at that time, his father described Anthony as "a piece of work." Anthony was a socially engaging child who was of low-average cognitive ability and was below grade level in a variety of academic skill areas, including reading, math, and written language. He also had sensory integration and language-processing difficulties and a variety of ADHD symptoms, including extremely poor organizational skills and excessive distractibility. Though he was a very concrete child, Anthony had a lot of trouble remembering things he'd learned. But perhaps the most striking thing about Anthony was his high level of anxiety. Anthony worried just about anything and everything. And there was one thing he was especially worried about.

"He's worried about the weather," his mother said in our first meeting. "If he hears it's going to snow, his day is ruined. Same thing with rain. It would be funny if it didn't happen so often and if he didn't become as upset as he does."

"Why's he so worried about the weather?" I asked.

"If it snows, he's afraid the snow will accumulate on the power lines and we'll lose power," the mother replied.

"Do you guys have a generator?" I asked.

"Of course," the mother said. "But he's afraid the generator will run out of gas."

"I'm sure you have extra gas for the generator," I said.

"Lots," Anthony's father replied. "But he's afraid we'll run out and won't be able to get more because gas stations can't pump gas without power."

"Well, he certainly has all his power-outage disasters covered," I said. "What's his deal with rain?"

"Lightening knocks out power," the mother replied. "Generator runs out of gas. Gas station can't pump more gas. Same deal. I've practically become a Weather Channel addict because I'm worried about him having a bad day! They do their five-day forecast and I'm on the edge of my seat!"

"Am I safe in assuming that you try to help him feel less anxious about these concerns?" I asked.

"Continuously," the mother said. "But on mornings when snow or rain is predicted, he's in his anxious mode and he's beyond reason. In fact, that's when he's most likely to have one of his meltdowns."

"How does that happen?" I asked.

"Well," the mother took a deep breath, "he gets himself very worked up over the weather and he gets going on the power outage thing and gets so upset and preoccupied that he can't be reasoned with. We try to reassure him, but that seems to make it worse. Eventually, he's crying and screaming and totally irrational. That about cover it?" she asked her husband.

"Oh, I think that sums it up pretty well," he replied. "He's really a nervous wreck. And we're at a loss for how to help him."

"And all because of the weather," I pondered aloud. "How long do the meltdowns last?"

"Well, eventually he runs out of gas," said the mother. "But at his worst, he can go on for several hours."

"How's he do in school—you know, on sunny days?" I asked.

"He gets a lot of academic help," the mother replied. "But he likes his teachers and he's always liked going to school. Less so, of course, on cloudy days."

"His teachers love him," the father said. "In fact, most adults do. But he's very concrete and not very good at taking other people's perspective. He's sort of oblivious to how he comes across. He's does have a few kids he plays with, though."

"He's really a very sweet kid," interjected the mother. "We

just haven't figured out how to help him with the weather thing."

Later in the session, I met alone with Anthony.

"I understand you have some concerns about the weather," I said.

"Yes," he replied. "When it snows, it ruins my whole day."

"Really? How come?"

"If there's too much snow, the power could go off."

"What would happen if the power went off?"

"It would be very bad. We'd be very cold. We couldn't cook our food."

"Hmmm. That *would* be bad. Is there anything *good* about when it snows?"

"Like what?" he asked, looking puzzled.

"I don't know," I replied. "It's kinda pretty."

"It's not so pretty when tree branches start falling on power lines."

"I guess you've got a point there. Does the power stay off for long?" I asked.

"I heard someone say it went off for three or four days once."

"That's a long time. What would your parents do if the power went off?"

"I don't know."

"Doesn't your house have a generator?" I asked.

"Yes."

"Couldn't your parents turn the generator on?"

"Yes. But it could run out of gas."

"Hmmm . . . I suppose it could at that. What would your parents do about that?" I asked.

"I don't know."

"Really? What do you think most people would do if their generator ran out of gas?"

"I don't know."

"Could they get more gas?"

"Yes. But the gas station can't pump gas if there's no power."

"What would your parents do about that?"

"I don't know."

The look on Anthony's face when he said "I don't know" was very convincing. What seemed clear was that Anthony's file of potential weather disasters was overflowing, but his file of what people do in the event of weather disasters so they don't keep worrying about it was empty.

His parents came back into the office.

"Isn't he great?" asked the father.

"So what are we going to do about the whole weather thing?" asked the mother.

"Well, to be perfectly honest, I'm not sure just yet," I replied. "I need to spend a little more time with Anthony to get a better sense of his way of thinking. I also need to get a better handle on why standard logic—you know, your reassurance and reasoning—aren't having the desired effect. And I'll need a better sense of the specific factors that lead to his meltdowns. But, to be honest, it seems like he lacks some pretty critical thinking skills. I suspect I'll end up trying to give you a road map for helping him think his way through his anxiety, in hopes that he'll eventually be able to use the road map on his own."

"You'd think all our reasoning would have rubbed off on him by now," the mother said.

"Indeed," I said. "It sounds like you've tried very hard."

"Is this his ADHD or his anxiety or his learning problems, or what?" asked the father.

"Well, my bet is that the whole package is coming into play in one way or another," I said.

"Is there any hope?" asked the mother.

"I'm always hopeful," I replied.

Danny

I first met with Danny and his mother when he was in the fifth grade. His mother and father had divorced amicably when Danny was seven and still considered themselves "co-parents." Danny and his younger sister stayed with the father and his fiancée every weekend. The mother described Danny as very bright (testing showed him to be in the superior range for general cognitive ability with no learning disabilities); perfectionistic; moody; irritable; and, as fate would have it, strikingly inflexible and easily frustrated. The mother was especially concerned about Danny's "rage attacks," which had occurred several times a week since Danny was a toddler. During such episodes, he would become verbally abusive and physically aggressive. These episodes were especially likely when things didn't go exactly as Danny had anticipated. He'd never had a rage attack at school. The mother was also worried about how these attacks were affecting Danny's sister, who, at times, seemed scared of her older brother and, at others, seemed to take some pleasure in instigating him.

Danny had seen numerous mental health professionals over the years; like many inflexible-explosive children, he'd accumulated a fairly impressive number of psychiatric diagnoses, including ADHD, oppositional-defiant disorder, obsessive-compulsive disorder (OCD), depression, and bipolar disorder. His family physician had medicated Danny with Ritalin several years previously, but Danny had remained moody, rigid, and explosive. The psychiatrist who diagnosed Danny with OCD had subsequently prescribed an antidepressant, but this medication caused Danny to become significantly more agitated and hyperactive.

"Danny can be in what seems to be a perfectly pleasant mood and then—bang!—something doesn't go quite the way he thought it would, and he's cursing and hitting," his mother reported. "I don't know what to do. The other day he and I were in the car together and I took a wrong turn. Danny suddenly became very agitated that it was taking us longer to get where we were going than it should have. All of a sudden, I had a ten-year-old kid punching me! In the car! While I was driving! It's insanity!

"I'm tired of people telling me this behavior is occurring because I'm a single parent. My ex-husband is still very much involved in Danny's life, and there hasn't been any of the backstabbing that occurs with some divorces. Anyway, these explosions started way before there were problems in our marriage, although, I must admit, he's a lot more explosive when he's with me than he is when he's with his father."

When I met with Danny, he seemed genuinely contrite over the behavior his mother had described. He said he'd been trying very hard not to be physically or verbally aggressive but didn't seem to be able to help himself.

At the beginning of one of our next sessions, the mother told me about Danny's biggest meltdown of the week. "Yesterday, I told him he had to come in from playing basketball to eat dinner. He whined a little, but I insisted. Next thing I know, his face is red, he's calling me every name in the book, he's accusing me of ruining his life, and I'm hiding behind a door trying to shield myself from getting kicked. I was petrified. So was his sister. And it's not the first time.

"Twenty minutes later, he was sorry. But this is just ridiculous," said the mother. "I'm sick of being hit, and it's just impossible to reason with him once he gets going."

"What did you do once he'd calmed down?" I asked.

"I punished him for swearing at me and trying to kick me," replied the mother. "I feel he needs to be disciplined for that kind of behavior."

"I can understand you feeling that way. Tell me, have you always punished him when he's acted like that?" I asked.

"You bet," the mother said. "I'm not willing to just let that kind of disrespect slide."

"What happens when you punish him?" I asked.

"He goes nuts," she said. "It's horrible."

"But despite all the punishing, he's still very verbally and physically aggressive, yes?"

"That's why I'm here." The mother smiled through gritted teeth.

"Well," I said, "I'm all in favor of punishment when it's productive—you know, when it's effective at changing a child's behavior. But I'm not real keen on punishment just for the sake of punishment."

"What, I should let him *get away* with what he does?" demanded the mother.

"Don't get me wrong," I said. "The behavior has to stop. But, based on what you've been telling me, 'not letting him get away with it' hasn't changed his behavior at all."

The mother pondered this observation for a moment.

"I think I figured that eventually the message would get through if I just kept plugging away," she explained. "I never stopped to think that maybe the message would never get through."

"Oh, I suspect Danny knows you don't like his behavior," I said. "In fact, I'm reasonably certain he even knows how you'd like him to behave."

"Then why doesn't he?" the mother demanded.

"Now that I've met with Danny a few times, I get the feeling he's generally in a pretty unhappy mood. I know he's not crazy about coming here, but is that his mood most of the time?" I asked.

"Absolutely," replied the mother. "We call him 'Grumpy.' He doesn't seem to enjoy himself very much . . . and he's very uptight. Everything seems to bother him."

"What an unpleasant existence," I said. "And it has very unpleasant implications for everyone around him."

"You can say that again," sighed the mother. "But what does that have to do with his being explosive and angry and trying to hurt me?"

"Well, if we view him as grumpy and irritable, rather than as disrespectful and oppositional, then I think our approach to dealing with him might be a lot different," I said.

"I don't understand what you mean," said the mother.

"What I mean is that kids who are grumpy and irritable often don't need more discipline," I said. "I've yet to see discipline be especially useful in helping a kid be less irritable and agitated."

"I still don't understand how being irritable is an excuse for his being so disrespectful and angry toward me," said the mother.

"Well, it's more of an explanation than an excuse," I replied. "But when people go through the day in an irritable, cranky mood, they experience every request or change or inconvenience as yet another demand for an expenditure of energy. If you think about it, over the course of a day or week, a person's energy for dealing with these requests and changes and inconveniences starts to wane. Often the event that sends an irritable, cranky person over the edge isn't necessarily the biggest; rather, it's the one that happened after he'd expended his last ounce of energy.

"Think of times you've been tired after a long day at work," I continued. "Those are probably the times when you're least adaptable and least flexible and when very minor things are likely to set you off. I think Danny is in that state fairly continuously."

"There's no way I'm going to tell him he's allowed to hit me just because he's irritable," said the mother.

"Oh, I'm not saying you should allow him to hit you," I said. "The hitting has got to stop. But to get the hitting and swearing and tantrums to stop, I think we need to focus on things you can do *before* he explodes, rather than on what you can do *after* he's explodes. And we need to focus on all the ingredients that are fueling his inflexibility and explosiveness. From what I've seen so far, helping him with his mood will be high on the list."

Eric

I first met with Eric and his parents when Eric was eleven years old. He was described as a bright, verbal, anxious, irritable, eleven year old who was already medicated with an antidepressant, which was being prescribed to help him be less anxious and moody. Previous testing had shown Eric's verbal skills to be markedly better developed than his nonverbal skills. His parents told me that at school, Eric had difficulty in math and with tasks that required writing. From what I was hearing, Eric's inflexibility-explosiveness seemed to be fueled by the combination of mood issues and a nonverbal learning disability. I came to know Eric as a remarkably literal, black-and-white thinker who was particularly inflexible around issues of morality.

During one of our early sessions, Eric's mother told me about an episode that had occurred the previous weekend. It seemed that Eric had been in Harvard Square with some of his friends. Harvard Square is an exciting, stimulating cluster of shops and interesting people near the Harvard campus. On weekends there are lots of entertainers—musicians, magicians, face-painters—and the area bustles with activity. Eric, his friends, and a few dozen other onlookers had stopped to watch a juggler perform. Every so often, an onlooker would drop money into the open guitar case laying on the ground next to the juggler. While the assembled people were enjoying the show, Eric noticed that the juggler didn't have a permit to be performing in Harvard Square (all performers in Harvard Square are supposed to obtain a permit from the police). Eric locked on to the notion that this was wrong. "All entertainers are supposed to have permits," he thought. "This juggler is breaking the law." During the next several minutes, Eric became increasingly agitated and less rational as he obsessed over the "crime" he was witnessing. He could think of only one thing to do: report the juggler to the police, which he promptly did. When Eric arrived with a policeman in tow, his friends were mortified (even the policeman seemed a little dumbfounded). "Why did you report the guy to the police?" the friends demanded as the policeman was telling the juggler the show was over. Eric was aware that his friends thought he'd done something off-base, but he couldn't figure out what they were so miffed about.

After his mother recounted this episode, Eric still seemed pretty clueless about why his friends were so upset with him. I decided to see if it was possible to get Eric to see the "gray" in a hypothetical situation.

"Eric, is it ever OK to speed in a car? Are there any situations in which it's acceptable to go over the speed limit?"

There was dead silence for the next five minutes. As Eric contemplated this question, I found myself thinking of the courageous but overwhelmed Tandy computer I had used in graduate school. Whenever I tried to pull up a file on this computer, it made promising "searching" noises while I sat on the edge of my seat and prayed. Most of the time, after much searching, my computer would ultimately find the file I'd requested. But about one out of every twenty times, after much searching, it would flash a message saying something to the effect of "Unable to locate file." I watched Eric search his brain's "files" and thought, "I wonder if the information he needs to answer this question is even in his head."

Finally, after a long search of his brain's "hard drive," Eric had at last pulled up a file. "It's OK to speed in a car if it's an emergency," he announced.

My expression didn't change, but inside I was jumping with glee. Eric *did* have the capacity to see the grays! The problem, of course, was that most real-life situations demand instantaneous flexibility. I thought, "If it takes him that long to think about this question in the safe, pleasant atmosphere of my office, he's in for some rough sledding when he's required to think quickly in Harvard Square." Still, it was a start.

In a later session, I witnessed one of Eric's meltdowns firsthand. Eric's parents wanted to discuss the possibility of Eric doing a few more chores around the house. As the discussion dragged on, Eric became increasingly agitated. He began to teeter on the edge of a meltdown, and I told him I thought it might be best for him to chill out for a while in the waiting area outside my office. He refused; unfortunately, he'd slipped slightly beyond the point of reason. His computer was crashing. His face turned red. He started yelling at his parents. He banged his chair into the ground. He stalked around the room. He was on the verge of tears. All because his parents wanted him to do one or two chores around the house. And then I noticed something fascinating: what he was saying was making no sense whatsoever.

"You guys do *not* have the right to be asking me to do these chores!" he shouted at his parents, sounding every bit like a spoiled brat. "I won't do them! I'll make your lives miserable! I hate you! Both of you! I always have! I'd like to do these chores! They're perfectly reasonable requests! You guys should be making me do these things! If you were good parents, you'd force me to do them!" Eric was becoming strikingly disorganized before our very eyes. He was becoming less and less rational.

"Eric, you're not making any sense," said the father. "Come on, calm down and let's talk about this rationally." But Eric was too far gone, and nothing any of us said got him back. Eventually, the meltdown ran out of gas and, exhausted, he started to cry.

Like Casey, Helen, and Anthony, Eric tended to have difficulty thinking clearly and became quite disorganized in the midst of frustration. And, like Helen, he was especially prone to frustration when his parents insisted that he shift from his agenda to their agenda. However, in his own distinctive way and for his own unique reasons, Eric was particularly inflexible over the specific issue of morality, and this inflexibility was very detrimental to his interactions with other people.

Mitchell

Mitchell was a fifteen-year-old ninth grader when I met him and his parents. I met first with his mother, a law professor, and father, a practicing lawyer, and was told that Mitchell was extremely unhappy about having been brought to my office that day, for he greatly distrusted mental health professionals. I was also told that Mitchell had been diagnosed with both Tourette's disorder and bipolar disorder, but was refusing all medication except an antihypertensive, which he was taking to control his tics. The parents reported that Mitchell, their youngest child (the others were already living away from home), was extremely bright and very eccentric, but was repeating the ninth grade because of a rough time he'd had at a local prep school the year before.

"This is a classic case of wasted potential," said the father. "We were devastated by what happened last year."

"What happened?" I asked.

"He just plain bombed out of prep school," said the father. "Here's a kid with an IQ in the 140s, and he's not making it at one of the area's top prep schools. He practically had a nervous breakdown over it. He had to be hospitalized for a week because he tried to slit his wrist."

"That sounds very serious and very scary. How is he now?" I asked.

"Lousy," said the mother. "He has no self-esteem left . . . he's lost all faith in himself. And he doesn't seem to be able to complete any schoolwork at all anymore. We think he's depressed."

"Where's he going to school now?" I asked.

"Our local high school," the mother replied. "They're very nice there and everything, but we don't think he's being challenged by the work, bright as he is."

"Of course, there's more to doing well in school besides smarts," I said. "Can I take a look at the testing you had done?"

I examined the psychoeducational evaluation that had been performed when Mitchell was in the seventh grade. The report documented a twenty-five-point discrepancy between his exceptional verbal skills and average nonverbal skills, difficulty on tasks sensitive to distractibility, very slow processing speed, and below-average written language skills. But the examiner had concluded that Mitchell had no difficulties that would interfere with his learning.

"This is an interesting report," I said.

"How's that?" asked the father.

"Well, it may give us some clues as to why Mitchell might be struggling to live up to everyone's expectations in school," I said.

"We were told he had no learning problems," the mother said.

"I think that was probably inaccurate," I said. I then explained the potential ramifications of some of the evaluation findings. As we talked, it became clear that Mitchell was indeed struggling most on tasks involving a lot of writing, problem solving, rapid processing, and sustained effort. "That's something we're going to have to take a closer look at," I said.

"Of course, he's still very bright," said the father.

"There are some areas in which he is clearly quite bright," I said. "And some areas that may be making it very hard for him to show how bright he is. My bet is that he finds that disparity quite frustrating."

"Oh, he's frustrated, all right," said the mother. "We all are."

After a while, I invited Mitchell to come into the office. He refused to meet with me alone, so his parents remained in the room.

"I'm sick of mental health professionals," Mitchell announced from the outset.

"How come?" I asked.

"Never had much use for them . . . none of them has ever done me any good," Mitchell answered.

"Don't be rude, Mitchell," his father intoned.

"SHUT UP, FATHER!" Mitchell boomed. "HE WASN'T TALKING TO YOU!"

The storm passed quickly. "It sounds like you've been through quite a bit in the past two years," I said.

"WHAT DID YOU TELL HIM!" Mitchell boomed at his parents.

"We told him about the trouble you had in prep school," the mother answered, "and about your being suicidal, and about how we don't . . . "

"ENOUGH!" Mitchell screamed. "I don't know this man from Adam, and you've already told him my life story! And I wouldn't have been suicidal if I hadn't been on about eighty-seven different medications at the time!"

"What were you taking back then?" I asked.

"I don't know," Mitchell said, rubbing hard on his forehead. "*You* tell him, mother!"

"I think he's been on about every psychiatric drug known to mankind," said the mother. "Lithium, Prozac . . . "

"STOP EXAGGERATING, MOTHER!" Mitchell boomed.

"Mitchell, don't be rude to your mother," said the father.

"If you don't stop telling me not to be rude, I'm leaving!" Mitchell screamed.

Once again, the storm quickly subsided.

"What medicines are you taking now?" I asked.

"Just something for my tics," Mitchell replied. "And don't even think about telling me to take something else! Let's just get off this topic!"

"He doesn't even take his tic medication all the time," said the mother. "That's why he still tics so much."

"MOTHER, STOP!" Mitchell boomed. "I don't care about the tics! Leave me alone about them!"

"It's just that . . ." the mother began speaking again.

"MOTHER, NO!" Mitchell boomed. His mother stopped.

"Mitchell, are you suicidal now?" I asked.

"NO! And if you ask me that again, I'm leaving!"

"He still doesn't feel very good about himself, though," the father said.

"I FEEL JUST FINE!" Mitchell boomed. "*You're* the ones who need a psychologist, not me!" Mitchell turned to me. "Can you do something about *them*?"

The father chuckled at this question.

"WHAT'S SO FUNNY?!" Mitchell boomed.

"If I could interrupt," I said, "I know you didn't want to be here today, and I can understand why you might not have much faith in yet another mental health professional. But I'm interested . . . what is it you'd like me to do about your parents?"

"Tell them to leave me alone," he growled. "I'm fine."

"Yes, he's got everything under complete control," the father said sarcastically.

"PLEASE!" Mitchell boomed.

"If I told them to leave you alone, do you think they would?" I asked.

"No," he glared at his parents. "I don't."

"Is it fair," I said, speaking carefully, "to say that your interactions with your parents are very frustrating for you?"

Mitchell turned to his parents. "You've found another genius," he said. "We need to pay money and waste our time on this guy telling us the obvious?"

"Mitchell!" said the father. "Don't be rude!"

"STOP TELLING ME WHAT TO DO!" Mitchell boomed.

"I appreciate your looking out for me," I said to the father. "But I actually want to hear what Mitchell has to say." I looked back at Mitchell. "I don't think I can get them to leave you alone without you being here."

"I don't think you can get them to leave me alone *with* me being here," Mitchell said. Then he paused for a moment. "How often do I have to come?" he asked.

"Well, to start, I'd like you to come every other week," I said. "I'd like your parents to come every week. Is that reasonable?"

"Fine!" he said. "Can we leave now?"

"I'd like to spend a few more minutes with your parents. But you can wait outside if you'd like." Mitchell left the office.

"Well, that's promising," said the mother. I couldn't tell if she was being sarcastic.

"We've got a lot of work to do," I said. "We've got a kid who's very easily frustrated, who's been diagnosed with Tourette's disorder and bipolar disorder, who tried to kill himself once . . . "

"Twice," interjected the mother. "He tried twice in the same year."

"Twice," I continued, "who's got very high expectations academically but some significant learning issues getting in the way, who isn't satisfactorily medicated at the moment, and who has no faith in the mental health profession."

The parents listened intently. "Where do we start?" the mother asked.

"Well, I need more information about a lot of things. But one thing is certain: We're not going to get anywhere unless I can establish a relationship with him. And we're going to have to make life at home a lot less tense and acrimonious so there's a least one domain where he's not frustrated to the max."

"Oh, would you say there's some tension in our family?" the father said sarcastically.

"A little bit." I smiled.

"So we come back next week?" asked the mother.

"You do," I said.

I hope that reading about these children gives you a better sense of what inflexibility-explosiveness can look like, how it can affect interactions with family members and other people, and how it "feels" for a child to be inflexible-explosive. These stories may also have provided a few hints about where this book is heading in terms of how to help a child be more flexible and handle frustration more adaptively. In the next chapter, I'll take a closer look at why the standard motivational approach that mental health professionals often recommend for reducing noncompliance and explosiveness in children may not be well matched to the needs of your child. Then, in later chapters, I'll talk about what you can do instead.

5
The Truth About Consequences

"You know, first we thought Amy was just a willful, spoiled kid," one father recalled. "Here were all these mental health professionals telling us that if we were simply firmer and more consistent with her, things would get better. Of course, Amy's grandparents added their two cents; they were constantly telling my wife and I that we were giving in too much. So we did the whole sticker chart and time-out routine for a long time. It makes me shudder to think of how much time that poor kid spent in time-out. But we were told the lessons we were trying to teach her would sink in eventually. Sometimes she wouldn't stay in the corner, and she'd try to kick us and bite us when we tried to hold her there. When we'd confine her to her room, she'd become destructive. Time-outs and rewards didn't seem to mean anything to her when she was upset. We couldn't figure out what we were doing wrong.

"So we went from doctor to doctor looking for answers. One doctor said Amy's tantrums were just her way of getting our attention and told us to ignore the tantrums and give her lots of attention for good behaviors. But ignoring her didn't help her calm down when she was frustrated about something. I don't care what the experts say, you can't just ignore your kid while she's being destructive and violent.

"Another doctor—this was around when she was eight years old—told us Amy had a lot of anger and rage. Amy spent the next year in play therapy, with this therapist trying to figure out what she was so angry about. He sort of ignored us when we told him Amy wasn't angry all the time, only when things didn't go exactly the way she thought they would. He never did uncover anything that accounted for her anger.

"The last person we went to told us Amy's tantrums were an expression of her 'emerging will and sense of self,' and we should let her express her emotions as much as possible. So we went through a few months of letting Amy 'express her sense of self' before my wife and I looked at each other and said, 'This is insanity!' In the meantime, we're still trying to figure out how to live with a kid like this.

"We've done everything we were told to do. We paid a big price—and I'm not just talking about money—listening to different professionals and trying strategies that weren't on target. And we became convinced that if the strategies weren't working, it must be our fault."

Psychology and psychiatry are imprecise sciences, and different mental health professionals have different theories and interpretations of inflexible-explosive behavior in children. As you now know, children may exhibit such behavior for any of a variety of reasons, so there's no right or wrong way to explain it and no one-size-fits-all approach to changing it. The key is to find explanations and interventions that are well matched to individual children and their families. In theory, this is a great idea. In practice, something often gets lost in the translation.

Probably the most recommended and widely used approach to understanding and changing the behavior of inflexible-explosive children is what is generically referred to as the standard behavior management, or motivational, approach. There are a few central beliefs associated with this approach. The first is that somewhere along the line, noncompliant children have learned that their tantrums, explosions, and destructiveness bring them attention or help them get their way by coercing (or convincing) their parents to "give in." This belief often gives rise to the notion that such behavior is planned, intentional, purposeful, and under the child's conscious control ("He's so manipulative! He knows exactly what buttons to push!"), which, in turn, often causes adults to take the behavior personally ("Why is he doing this to me?"). A corollary to the belief that such behavior is learned is that the child has been poorly taught ("What that kid needs is parents who are willing to give him a good kick in the pants"). Parents who become convinced of this belief often blame themselves for their child's inflexible-explosive behavior ("We must be doing something wrong . . . nothing we do seems to work with this kid"). Finally, if you believe that such behavior is learned and the result of poor teaching, then it follows that it can also be unlearned with better and more convincing teaching.

In general, this reteaching and unlearning process includes (1) providing the child with lots of positive attention to reduce the desirability of negative attention; (2) teaching parents to issue fewer and clearer commands; (3) teaching the child that compliance is expected and enforced on all parental commands and that he must comply quickly because his parents are only going to issue a command once or twice; (4) delivering consequences—rewards, such as allowance money and special privileges, and punishments, such as time-outs and the loss of privileges—contingent upon the child's successful or unsuccessful fulfillment of specific target behaviors (such as complying with adults' commands, doing homework, and getting ready for school); and (5) teaching the child that his parents won't back down in the face of tantrums. This approach isn't magic; it merely formalizes practices that have always been important cornerstones of effective parenting: being clear about how a child should and should not behave, consistently insisting on appropriate behavior, and motivating the child to perform such behavior.

Some parents and their children benefit enormously from such formality, find that these procedures give some needed structure and organization to family discipline, and end up sticking with the program for a long time. Other parents benefit mostly from what they learn in behavior management books or courses about their counterproductive parenting habits. Such parents may not stick with a formal behavior management program for long but still change some fundamental aspects of their approach to parenting and therefore become more effective at teaching and motivating their children. But for other parents, implementing a behavior management program is a significant departure from their natural style or philosophy of parenting. So, although they understand the logic of behavior management procedures, they neither stick with the program nor make fundamental changes in their parenting behavior. Many such parents may embark on a behavior management program with an initial burst of enthusiasm, energy, and vigilance, but become less enthusiastic, energetic, and vigilant over time. They often return to their old ways of doing things, and their child's tantrums and outbursts eventually return as well. Thus, one reason behavior management procedures may not be effective for some children is that their parents fail to implement the procedures correctly or consistently.

But there's another reason, and the second explanation is central to the rationale for this book: Children who are developmentally compromised in the skills of flexibility and frustration tolerance may lack the capacity to shift immediately and consistently from their agenda to their parents' agenda, even when faced with meaningful consequences. And none of us can consistently exhibit behaviors of which we are incapable, no matter how enticing the reward or how aversive the punishment. Motivational strategies don't make the impossible possible; they make the possible more possible.

Motivational strategies don't make the impossible possible; they make the possible more possible.

In other words, if the Chicago Bulls offered me Michael Jordan's salary to play basketball like Michael Jordan, I would be extremely motivated (yes, I could probably think of a few things to do with $30 million), but I would still have mediocre basketball skills. Oh, maybe I'd be able to sink an occasional great shot, but—and you'll have to take my word on this, having never seen me play basketball—that definitely wouldn't mean I'd be able to sink great shots consistently. If I was then punished for failing to play basketball like Michael Jordan, I'd start to become frustrated and would probably begin to wonder what was the matter with me (and with the people who had unrealistic expectations of me).

Consequences can be very useful in two specific ways: as an instrument of *learning* or as an instrument of *motivation*. But the vast majority of the inflexible-explosive children with whom I work are well aware of the behaviors adults like and dislike, so consequences aren't necessary for the purpose of further teaching; the child already knows. And—hold on to your hat—the vast majority of inflexible-explosive children with whom I work are already motivated not to make themselves and those around them miserable, so consequences aren't necessary for the purpose of additional motivation. By definition, a consequence is an event that occurs following a behavior—in other words, on the back end. For a consequence to achieve its desired effect, you have to have faith that something you do on the back end is going to be accessible and meaningful to the child the next time he's frustrated. So what do we make of a child who already *knows* inflexibility and explosiveness are undesirable and is already *motivated* to be less inflexible and explosive, or whatever skills they do have are inaccessible to them in the midst of frustration? That if the child *could* be less inflexible and explosive, he most assuredly *would*.

For a consequence to achieve its desired effect, you have to have faith that something you do on the back end is going to be accessible and meaningful to the child the next time he's frustrated.

Let's go back and take a closer look at what happened when Amy's parents were instructed on how to implement behavior management procedures. First, the parents were taught how to give directions in a way that made it easier for Amy to comply and were encouraged to "catch Amy being good" (with verbal praise, hugs, and the like) every time she complied. The next week, she and her parents were asked to identify a variety of meaningful rewards that could be earned in exchange for compliance, and the parents were helped to design a "currency" system—in her case, a point system—as a way of keeping track of the percentage of times Amy complied with their requests. The points were to be exchanged periodically for the rewards, each of which had a price tag. The next week, the parents were taught the proper way of implementing the time-out procedure when Amy did not comply. The training was now complete. By the third week, Amy was receiving a specified number of points every time she complied with a parental request and was confined to time-out and lost points when she did not comply. Amy was now, most assuredly, very motivated to comply (assuming, of course, that she wasn't motivated in the first place).

The following scenario ensued. The parents would give directions. Amy would become frustrated. The parents would remind her of the consequences for not obeying and of the necessity for immediate compliance. Rather than helping Amy immediately gain access to the file in her brain that contained the critical information ("If you do what they're asking, you'll earn points; if you don't, you'll get a time-out"), her parents' warning would actually cause Amy to become *more* frustrated and agitated, her thinking increasingly disorganized and irrational, and her control over her words and actions greatly reduced. In other words, vapor lock would commence. In accordance with their training, Amy's parents would interpret her increased intensity and failure to respond to their commands as an example of coercive noncompliance and would warn her of an impending time-out. The result: meltdown.

Amy, now bereft of any semblance of coherence or rationality, would begin screaming and lashing out. Her parents, safe in the belief that the tantrum was merely an attempt to manipulate them into giving in, would stick with their training and try to get Amy into time-out, an action that would intensify her meltdown and deepen her incoherence. Amy would resist being placed in time-out. Her parents would try to restrain her physically in time-out (many therapists no longer recommend this practice, but Amy's therapist wasn't one of them) or confine her to her room until she calmed down. The struggle to keep Amy in time-out or confine her to her room would further intensify her meltdown. She would try to hit, kick, bite, scratch, and spit on her parents. Once in her room—when her parents were actually able to get her there and keep her there—she would try to destroy whatever "destructibles" were handy.

Eventually, meaning somewhere in the range of ten minutes to two hours, Amy would become completely exhausted and start to cry or go to sleep. Coherence would be restored. Her exhausted parents would be frustrated and angry and would hope that what they just did to their daughter—and endured themselves—was eventually going to pay off in the form of improved compliance. When Amy would finally emerge from her room, she would be remorseful. The parents, once again in accordance with their training, would, in a firm tone, reissue the direction that started the whole episode in the first place.

In subsequent sessions with their therapist, the parents were commended for their efforts and reassured that Amy would eventually learn that she had to comply with their requests. They were reminded that Amy's destructive, assaultive behavior was an attempt to coerce them into relenting and were encouraged not to give in. They were told that their daughter's ultimate remorse was an attempt to manipulate them. The therapist recommended that they stick with the program. For months. And months. Even though things weren't improving appreciably.

What's the matter with this picture? Amy lacked two critical ingredients necessary for the success of the motivational program:

1: She was apparently not *capable* of quickly shifting gears—from Agenda A (her agenda) to Agenda B (her parents' agenda)—in response to her parents' commands.

2: She was apparently not *capable* of remaining relatively calm, coherent, and organized when frustrated by her parents' commands, so she could gain access to her knowledge that (a) compliance with parents' commands is what children are supposed to do; (b) if she complied, she would be rewarded; and (c) if she did not comply, she would be placed in time-out. Consequences can be effective if a child is in a state of mind to appreciate their meaning, but they don't work nearly so well if a child is not able to maintain such a state of mind.

Consequences can be effective if a child is in a state of mind to appreciate their meaning, but they don't work nearly so well if a child is not able to maintain such a state of mind.

Her parents, on the advice of a well-intentioned mental health professional, had conceived of Amy's difficulties as motivational in nature and thus were trying to motivate their daughter to exhibit behaviors of which she was incapable. Given the multiple pathways to inflexible-explosive behavior, was poor motivation really the best explanation for Amy's difficulties? Was Amy's noncompliance truly planned, purposeful, intentional, and under her conscious control? Are the terms *oppositional, noncompliant, manipulative, coercive*, and so forth really the best ways to describe Amy? Are her parents truly lousy teachers? Is more and better discipline the answer to a difficult temperament? To difficulties related to mood or anxiety? To deficits in executive functions, social skills, language processing, nonverbal skills, and sensory integration?

This seems like a good time to point out that there are some less familiar but important considerations emanating from the more contemporary variants of the behavior management approach that I believe are critical to the process of helping a child be less inflexible and explosive. These considerations include being cognizant of the events, called "antecedents," that precede or fuel inflexible-explosive episodes; being aware of situational factors that may fuel such episodes; paying attention to what's going on (or not going on) inside a child's head that may be related to such episodes; engaging in progressive training of various skills that such a child may be lacking; and being aware of the possible neurobiochemical underpinnings of such behavior. As you'll see, these often-neglected principles are an important facet of the alternative approach described in this book.

What I've learned over the years is that some children become so overwhelmed so quickly by their frustration that their capacity to maintain coherence in the midst of frustration is severely compromised. I've seen that these children are also compromised in their ability to gain access to information they've stored from previous experiences and think things through so as to formulate a well-organized, reasoned response to frustrating situations. And, as you've read, I've come to believe that the inflexibility and explosiveness evidenced by these children often have a neurobiochemical basis.

Meaningful, immediate, consistent consequences may *not* be "just what the doctor ordered" for these children.

Your child is drowning in a sea of frustration and inflexibility. You are the lifeguard. If you swim out to your drowning child, hold up a dollar bill, and yell "SWIM!" your child will continue to drown. If you ignore your child, he'll just as surely drown. Your days of trying to motivate your drowning child to swim are over. If your child could swim, he would. It's time for Plan B.

"We're encouraging people to become involved in their own rescue."

Question: "Aren't flexibility and frustration tolerance critical skills? Doesn't my child have *to change?"*

Answer: "Flexibility and frustration tolerance are critical skills, and there may be some ways to teach your child how to be more flexible. But you may not be getting anywhere—or doing any productive teaching—by engaging in frequent battles with him whenever you try to force him to be more flexible."

Question: "But if I don't teach my child how to be flexible, how will he learn?"

Answer: "If he's going to learn to be more flexible—and I'm optimistic that he can—it's not going to happen by your being a role model for inflexibility."

Question: "But it worked for me; I'm just raising my kids the same way I was raised."

Answer: "The way you were raised may have worked for you— and it seems to be working for your other children—but it's clearly not working as well for your inflexible-explosive child."

Question: "Don't I need to set a precedent now so my child knows who's boss and doesn't think he can always get his way?"

Answer: "Your inflexible-explosive child probably already knows you are the boss. During vapor lock, your child is in a different sphere, and the question of 'who's the boss?' doesn't really enter his 'thoughts.' In fact, if you're trying too hard to show your child who's the boss, you may actually be fueling his meltdowns and setting the stage for an adversarial pattern of parent-child interactions. And my sense is that little or no learning takes place for your child during meltdowns because he's basically incoherent.

Teaching a child who's boss is easy; understanding and responding to a child's deficits in the areas of flexibility and frustration tolerance is a lot harder."

Question: "But aren't most kids a lot more flexible than mine?"

Answer: "Bingo."

❄

Marvin's mother began to cry. "I can't take this anymore," she said, wiping her eyes. Her husband looked on helplessly, holding back tears of his own. "I'm so worried about him. . . . Marvin can be so wonderful sometimes . . . but the minute something doesn't go his way, he goes crazy on us. And I'm tired of getting hit by an eight year old."

You met Marvin—a child diagnosed with ADHD, depression, and Tourette's disorder—in Chapter 1. His parents had come to the conclusion that rewards and punishments were not the answer to Marvin's inflexibility and explosiveness, but they had continued trying to "motivate" compliant behavior because they couldn't think of anything else to do.

"Are we all on the same page here about Marvin's behavior?" I asked in one early session.

"What do you mean?" asked his mother.

"Are we all convinced that Marvin isn't doing what he's doing for attention, to be manipulative, to coerce you into giving in, or to yank your chains?" I asked. "In other words, are we ready to leave the motivational explanation for his difficulties behind us?"

"Absolutely," said the father. "You've helped us understand that there are a lot of different factors—his chronic irritability, inflexibility, impulsivity, tics—that are making it very hard for him to deal well with frustration."

"But you're still using strategies that are intended to motivate compliance," I said. "If we don't think his problem is motivational, I'm not sure why we'd think motivational strategies would be the best way to help him."

"So I'm not supposed to punish him when he hits me?" asked the mother.

"Do you have any faith that punishment makes it any less likely that he'll refrain from hitting you the next time he's frustrated?" I asked.

She thought for a moment. "No," she replied. "Punishing him certainly hasn't gotten us anywhere so far."

"That's for sure," said the father. "In fact, it's probably made things worse in some ways."

"Punishment is a back-end response to Marvin's difficulties," I said. "It's something you do once he's already fallen apart. But I don't have a lot of faith that anything you do on the back end is going to make much difference. In fact, as best as I can tell, it's merely made everybody more frustrated and produced a lot of adversarial interactions. I think we need to focus our efforts on the front end to try to keep Marvin from falling apart in the first place."

"I don't understand," said the mother. "It sounds like you're telling us not to respond at all."

"Oh, I want you to respond all right," I said. "But at different points in time—in other words, before and during vapor lock and at the crossroads, rather than before and after meltdowns—and in a very different way. In fact, the things I think you should be doing are a lot harder than simply responding to Marvin's meltdowns with punishments."

"What should we be trying to do?" asked the father.

"Our first goal is to do a better job of anticipating situations that will be frustrating to Marvin and to be much more judicious about the frustrations we choose for him to deal with," I said. "Our second goal is to help Marvin stay coherent when he gets frustrated so he can think rationally. If we can keep him coherent, we may be able to start working on our third goal, which is to help him think things through when he becomes frustrated, so he doesn't end up hitting you and calling you names. Rather than punishing him and teaching him who's boss, you'll instead be trying to help him with some of the things he's really struggling with."

6
Clear the Smoke

By now you should have a much better sense of why it's crucial for you to understand your child's inflexibility-explosiveness, the factors that may contribute to it, and why an approach aimed solely at motivating compliant behavior may not be well suited to your child's needs. Of course, you still don't have the details on the most important issues: What should you do instead? How do we help a child who—because of a difficult temperament, deficits in executive skills, hyperactivity and impulsivity, a depressed or unstable mood, anxiety, language or nonverbal impairments, or deficits in social skills—lacks the capacity to respond adaptively to life's demands for flexibility and frustration tolerance? What might we do differently? As it turns out, there are some things that are quite helpful to many of the children, parents, and teachers with whom I work.

The approach I encourage is comprised of two basic avenues. The first avenue involves the creation of what I call a "user-friendlier environment"—an environment in which your child's deficits in the areas of flexibility and frustration tolerance are less of a handicap. A user-friendlier environment can help clear the smoke and set the stage for you and your child to begin working together to deal more effectively with his inflexibility and explosiveness. In coming to a clearer understanding of your child's difficulties, you've already taken the most important step down this avenue. By demonstrating to your child that you understand how debilitated he becomes in situations that require flexibility and a tolerance for frustration, you'll help him maintain coherence in the midst of these situations so he can think through and discuss potential solutions. A user-friendlier environment is also intended to help you (1) be more realistic about which frustrations your child can actually handle and more open to eliminating unimportant, unnecessary frustrations, thereby greatly reducing the opportunities for vapor lock and meltdowns; (2) think more clearly when your child is in the midst of vapor lock and understand that *you*—not your child—determine whether a situation develops into a full-fledged meltdown; (3) be more attuned to the situations that routinely cause your child the greatest frustration; (4) move away from an adversarial relationship with your child while maintaining your role as an authority figure; and (5) not take your child's inflexible-explosive episodes so personally. In other words, creating a user-friendlier environment should contribute to some fundamental changes in your relationship with your child. This chapter—and several that follow—elaborate on the user-friendlier theme (this theme is extended to the school environment in Chapter 11).

By demonstrating to your child that you understand how debilitated he becomes in situations that require flexibility and a tolerance for frustration, you'll help him maintain coherence in the midst of these situations so he can think through and discuss potential solutions.

The second avenue involves addressing your child's difficulties in a more direct way so as to improve his capacity for flexibility and frustration tolerance so he doesn't require a user-friendlier environment forever. This is discussed in Chapters 9 and 10.

Picture this: You're walking along and you come upon a man in a wheelchair whose path has been blocked by a flight of stairs. If you're interested in helping the man get past this obstacle, you've basically got two choices: (1) insist that he get out of the wheelchair and walk up the stairs, or (2) find a way to help him circumnavigate the stairs (for instance, help him find a wheelchair ramp). The first choice is implausible because it fails to take into account the man's limitations. No matter how motivated he may be, his physical condition still imposes limitations that prevent him from getting out of the wheelchair and walking up the stairs. If you're one of those fortunate souls for whom walking up stairs comes naturally, it may be hard to appreciate the struggles of a person who has difficulty doing so. Still, you probably don't have any trouble appreciating the importance of making the environment "user friendlier" for wheelchair-bound individuals by providing wheelchair ramps, rest-room accommodations, and preferential parking.

Now consider this: Because of any (or many) of the factors described in Chapter 3, your child has limitations that confine him to a *cognitive* wheelchair in situations requiring flexibility and frustration tolerance. The reason your child has been melting down so often is that his path is chronically being blocked by the normal demands for flexibility and frustration tolerance that are ever present in the environment. In the early stages of trying to help your child, you've got two choices: (1) to continue to insist that he be more flexible and handle frustration more adaptively or (2) to find ways to reduce the demands for flexibility and frustration tolerance in situations he can't seem to steer his way around—in other words, find wheelchair ramps. As you've no doubt discovered, the first choice often doesn't work, no matter how appealing the incentives or aversive the punishments you may apply to enhance motivation. A user-friendlier environment has more to do with the second choice.

If you're one of those fortunate souls for whom flexibility and frustration tolerance come naturally, it may be hard to appreciate fully the struggles of a person who has difficulty in these domains. And in one respect, it's to your child's great misfortune that he looks "normal" (people tend to be a lot more compassionate when a child's difficulties are clearly apparent and not subject to a variety of interpretations). So, with all the labels that are likely to be thrown at you to explain why your child is inflexible and explosive, it's critical for you to keep your eye on the ball: Your child has developmental deficits in the skills of flexibility and frustration tolerance.

Many parents accept the idea of inflexibility-explosiveness but want to jump right into the second avenue of intervention: addressing their child's difficulties directly. Unfortunately, the direct approach is usually not the ideal place to start. Few of us learn best in an environment we experience as tense, hostile, and adversarial, and your child is no exception. So the first avenue is the optimal starting point because this first pathway actually sets the stage for helping your child become increasingly receptive to new learning. Yes, this is a departure from the "throw 'em right into the water" approach, but it seems to work much better for a lot of children.

What type of environment would work better for a child compromised in the skills of flexibility and frustration tolerance? If you reflect on the adjustments we're willing to make for people with other handicapping conditions, you'll get an idea about where we're heading. People who are visually impaired often receive assistance from a seeing-eye dog. Children with reading disabilities often receive extra help. And people who need extra time boarding an airplane often get extra time to board.

By way of reminder, the components of a user-friendlier environment described next share a critical common ingredient: They emphasize responding to your child *before* he's at his worst, rather than when or after he's at his worst. As you know, I have little faith that back-end interventions will have much impact on your child the next time he becomes frustrated. In other words, the best thing that adults can do about meltdowns is to make sure they don't happen in the first place.

- **A user-friendlier environment is one in which all the adults who interact with the child have a clear understanding of his unique difficulties, including the specific factors that fuel his inflexibility-explosiveness.**

If you're a child who has trouble thinking through different options for dealing with frustration or who rapidly becomes overwhelmed by the emotions associated with frustration, it can be calming to be in the presence of an adult who knows you well, has a clear sense of what's going on, and feels empowered to act as a temporary guide to steer your cognitive wheelchair around a frustrating episode. As you've already read, when children are stuck in the red haze of inflexibility and frustration, they respond a lot better if they perceive adults as potential helpers, rather than as enemies. In contrast, adults who don't have a clear understanding of what's going on—or who have unenlightened, unrealistic expectations—may unwittingly place additional obstacles in the child's path.

One child with whom I was working had been melting down far less frequently at home, and his parents and I had started thinking we were on easy street—until his physical education teacher, whom we had neglected to enlighten, demanded that the child wear a sweatshirt outside on a fifty-five-degree day. After about five minutes of what might best be called "reciprocal inflexibility," the child put his fist through a window. The point here is that it's user friendlier for moms and dads and principals and teachers and coaches to be on the same wavelength and to have a common set of expectations, so a child who has trouble adapting quickly in the first place doesn't have to figure out which wavelength and expectations are in effect at any given moment. Moreover, achieving an in-depth, accurate understanding of the precise factors underlying a child's difficulties is critical to designing interventions that are well matched to his needs.

When children are stuck in the red haze of inflexibility and frustration, they respond a lot better if they perceive adults as potential helpers, rather than as enemies.

- **In a user-friendlier environment, parenting goals must be judiciously prioritized, with an emphasis on reducing the overall demands for flexibility and frustration tolerance being placed on the inflexible-explosive child.**

The theory here is that if you eliminate many of the unnecessary obstacles in the path of your child's cognitive wheelchair—in other words, reduce the opportunities for vapor lock and meltdowns—your child's global level of frustration should decrease, and he should melt down less readily in response to the obstacles that remain. Another user-friendlier development: Your own global level of frustration should decrease correspondingly. Also, if your child is melting down less often, the general level of tension and hostility in your family should diminish, thereby facilitating improved parent-child interactions. These improved interactions can set the stage for—hold on to your hat again—actual discussions in the midst of frustrating situations instead of vapor lock and meltdowns. But we're not there yet.

While many parents and teachers can appreciate the wisdom of reducing the overall demands for flexibility and frustration tolerance being placed on their child, they often need help understanding how to do it. They want reassurance that the child will not come to view them as "pushovers." Don't forget: Most inflexible-explosive kids are quite clear about who's the boss when they're in a coherent state. So Job 1 is to make sure they stay coherent even when they are frustrated, and a little judicious prioritizing of when and on what issues adult authority is asserted can go a long way. Chapter 7 presents a thorough description of what I've come to call "basket thinking," a modified form of picking your battles, so I won't elaborate on the mechanics of prioritizing here.

- **A user-friendlier environment is also one in which adults try to identify—*in advance*—specific situations that may routinely lead to inflexible-explosive episodes.**

There's little question that some episodes seem to come out of the blue, but many are actually predictable. And if they're predictable, you'll be taken by surprise a lot less often and can try to have the wheelchair ramp ready in advance or avoid the situation altogether. Just don't expect preparation to work all the time because it won't.

Some parents find it useful to keep an informal record of their child's meltdowns for a few consecutive weeks to try to determine whether there's some sort of predictable pattern. What types of situations seem to contribute to vapor lock and meltdowns in your child? Waking up and getting out of bed in the morning? Getting ready for school? Doing homework? At bedtime? Boredom? Shifting from one activity to another? Being in the company of certain individuals? Being hungry just before dinner? Being surprised by a sudden change in plans or in some aspect of the environment? When a social interaction requires an appreciation of the gray in a situation? When a situation has become too complex for him to sort through? When his Ritalin wears off? Are there certain topics, sounds, or articles of clothing that seem to induce meltdowns fairly regularly? Which of these situations can we alter? Which ones can we prepare for in advance? What situations can we avoid altogether, so as to reduce the child's frustration? Which ones can't be altered, prepared for, or avoided? We've got a lot to learn about your child! I discuss this more in Chapter 9.

- **In a user-friendlier environment, adults read the warning signals and take quick action when these signals are present.**

As you've already read, it's the rare child who, as frustration is kicking in, states in a calm, coherent manner, "Uh, folks, this situation is beginning to cause me intense frustration, and I'm reasonably certain that my capacity for rational thought is rapidly diminishing. So, I'm wondering if we might begin to consider some alternative course of action, so I don't lose it completely." Most of the time, the child provides a less articulate but still meaningful indication that vapor lock has commenced and an altered state of mind has set in. Some children will say, "I'm tired" or "I'm hungry"; others will say, "I can't do this right now!" or "I'm bored" or "I don't feel right" or, in the case of a child you already know, "I was going to have those three waffles tomorrow morning!" Still others may blurt out "I hate you!" or "Fuck you!" Others won't say anything but their nonverbals—body language, sudden crankiness, whining, irritability, restlessness, a sudden loss of energy for handling routine tasks or activities (even ones that are normally pleasurable) or an unexpectedly intense response to routine requests—are a dead giveaway. These are the signs that your child is moving toward the edge.

The burden for recognizing these signs falls on the adults who are interacting with the child. It's critical to read these early warning signs and take action while there's still some coherence to work with to prevent these hints from turning into something much bigger and uglier. A former lifeguard I know, Brian Blanchard, put it this way: "Lifeguards who watch the water closely don't have to jump in and rescue people very often."

What should you do if you read some of the early warning signs? If it's very early in a vapor-lock sequence, there's a chance that the combination of empathy and logical persuasion will still work.

Mother [*calling upstairs*]: Addie, it's time to go to the picnic.

Addie [*no response*].

Mother [*walking upstairs*]: Come on, Addie, this is our big family reunion; there's going to be lots of people there, and I don't want to be late.

Addie [*from under the bedcovers*]: I'm not going.

Mother [*at the door of Addie's bedroom*]: Addie, why are you back in bed? It's almost one o'clock.

Addie [*hiding her face*]: I don't want to go.

Mother [*thinking to herself*]:

A. We're in vapor lock; my next response is critical.

B. I wonder what's going on with Addie?

C. I'd better take some time to find out, even if it means being a little late to the reunion.

D. What is it about going to the reunion that may be causing Addie to behave this way?

E. Well, she's not very good at dealing with big crowds, she won't know many people there, and she's not very good at organizing an adaptive response to things that are making her uncomfortable.

Mother [*gently*]: Addie, how come you don't want to go to the reunion?

Addie: I don't know; I just don't want to.

Mother: Hmm . . . let's think about this; there are going to be a lot of people there, and you won't know that many of them.

Addie: And I don't know who I'll play with.

Mother: I was thinking that might be troubling you. Let's think about this a second. There are some people you will know— Uncle Dave, Aunt Julie, cousins Fay and Debbie—you always like playing with cousin Fay.

Addie: Cousin Fay will be there?

Mother: She's probably there already. But I'll tell you what. If you want, for the first few minutes we're there, why don't me and you stick together until you find someone you want to play with.

Addie: OK.

Empathy can help a child hang in there, and a coherent child has a much better chance of responding to logical persuasion than does an incoherent one. Nonetheless, for severely inflexible-explosive children, I tend to think of empathy and logical persuasion as a fairly low-odds intervention, at least early on; after a few months of living in a user-friendlier environment, this approach has a better chance.

Aside from empathy and logic, there are a few other quick interventions that may also be useful early in vapor lock. One of my favorites is distraction. With distraction, you're redirecting your child to some other activity so the inflexibility and frustration involved in the current activity are diminished. In my experience, distractions should be enjoyable to your child and require minimal brain energy (thinking and processing) from him. Distractions that meet these criteria have a chance. Of course, humor is probably my favorite distraction. An example of a ingenious distraction is an innovative product on the market called Magic Cream. To parents of regular children, Magic Cream is nothing more than a simple hand cream with sparkles in it. To parents of inflexible-explosive children, the description on the label sounds like nirvana: "Designed to relieve tantrums, whining, crankiness, and laziness." Younger children who are on the verge of meltdowns are sometimes able to shift gears rapidly when their parents suggest the quick application of a little Magic Cream. Other distractions can work equally well.

Another quick intervention is what I call "downshifting." In the same way that you wouldn't downshift from fifth gear (Agenda A) to reverse (Agenda B) in one easy step in a car (unless you wanted to tear up your transmission), you shouldn't try the same thing with an inflexible-explosive child (unless you want to tear up his). What you should do instead is to try to shift the car (child) slowly from fifth gear to fourth, fourth to third, third to second, and second to first; put on the brakes; and finally put the car (child) into reverse—in gradual steps. All the while, listen closely to the sound of the car's (child's) "engine," so you can maintain a good feel for how close you are to having the transmission blow.

Continuing with the car theme, the following story was told to me by the mother of a very concrete, language-impaired ten-year-old girl with whom I'd begun working:

❄

Caroline's parents had a few family friends over to their house for a get-together. At one point, Caroline's mother and some of her friends decided to go for a drive to the shopping mall, and Caroline was excited about being asked to go with them. Caroline eagerly ran over to her father and asked him for the keys to the car. The father, who on occasion permitted Caroline to steer the car while seated in his lap, was engaged in conversation with one of the friends, and, with barely a second thought, gave Caroline the keys to the car.

When the mother and her friends came out to the car, Caroline, smiling and firmly planted in the driver's seat, proclaimed, "I'm driving!"

The friends and the mother laughed, thinking that this behavior was adorable. Which, of course, it would have been had Caroline not been dead serious.

"OK, honey, slide over so I can drive," said her mother.

"I'm driving!" insisted Caroline, her smile a bit less certain.

The mother recognized that vapor lock had commenced, but her embarrassment and desire to get to the mall kept her from reacting as she later wished she would have.

"Caroline, move over!" the mother demanded, moving her daughter closer to the edge of a full-blown meltdown (but no closer to the passenger seat).

"I'm driving and I'm not moving over!" insisted Caroline, no longer smiling. The mother's friends began to murmur. Not good.

"Caroline, I'm going to count to three, and then you'd better be in the passenger seat!" insisted the mother.

Commence meltdown. A scene ensued in which the mother tried to drag Caroline from the car and Caroline clung to the steering wheel for dear life. Eventually, the father joined the commotion and was able to pry Caroline loose. Caroline was subsequently confined to her room, where she spent the next forty-five minutes screaming and destroying her belongings. The guests left (still murmuring). The mother and her friends never made it to the shopping mall.

Downshifting might have had a chance with Caroline in this situation. In our next session, the parents assured me that Caroline's legs didn't come close to reaching the gas pedals and that she didn't know how to get the key into the ignition, so safety had not been the mother's primary concern. The mother felt she'd overreacted because of the time pressure involved, but agreed that enduring forty-five minutes of Caroline's screaming and throwing toys was far more unpleasant and far less efficient than spending a few minutes helping her daughter downshift.

"The first thing I should have done is tell her how great it was that she'd gotten the car ready for us. That probably would have gotten her from fifth gear to fourth gear. She's usually more receptive when we empathize with her first. Then I should have asked her to show us how she would drive the car and to show us the different equipment she would use. That probably would have gotten her from fourth gear to third. Then I could have told her that she could sit in my lap and help me steer the car out of the driveway—that might have gotten us from third gear to second. I suppose once I had let her get us out of the driveway on my lap, she would have slid right over to the passenger seat and we would have gone on our merry way."

The father tried to comfort his wife. "Well, I didn't exactly jump in and act like 'Mr. Reasonable' either. I think we both should have been more worried about understanding what was going on in Caroline's head than about how our friends felt."

Empathy and logical persuasion, distraction and humor, and downshifting may be useful early in the vapor-lock sequence, but they may be useless if a child's incoherence deepens, thereby requiring a quick shift to the basket thinking described in the next chapter. What this means, of course, is that reading your child must continue past the point of merely recognizing the onset of vapor lock; it's important to continue reading to gauge the effect of your attempts to help him stay coherent.

By the way, one reason why ignoring isn't particularly effective for these children during vapor lock and meltdowns is that it doesn't provide them with the road map they need to help them find safe passage out of their frustrated state. What they need to help them find their way is a coherent adult.

- **In a user-friendlier environment, adults can interpret incoherent behaviors for what they really are: incoherent behaviors.**

Accurately interpreting incoherent behaviors is very hard to do. Nonetheless, during meltdowns, all bets are off on what kind of mental debris may come out of a child's mouth. When inflexible-explosive kids scream "Fuck you!" or "I hate you!"

or "I wish I was dead!" in the midst of vapor lock or meltdowns, they often mean "My capacity for rational thought is rapidly diminishing or is already completely gone!" If you immediately punish the child for saying "Fuck you!" because it's offensive and disrespectful— which it certainly is—you've taught another lesson on the importance of respect (a principle to which your child may have no trouble adhering when he's coherent), but you've also thrown fuel on the meltdown fire, all of which may prove fairly meaningless the next time the child is frustrated. If you read the "Fuck you!" as "My capacity for rational thought is rapidly diminishing" and use any of the various strategies described in this and the next three chapters to try to help steer the child around or through his frustration, you'll probably avoid a full-fledged meltdown and may get in some good teaching, role modeling, and relationship building. It's your choice.

- **In a user-friendlier environment, adults can also understand the manner in which they may be fueling a child's inflexibility-explosiveness.**

Adults may dwell on certain issues so long or so often that they practically invite meltdowns. They may try so hard to be a good influence that they try to change everything about a child all at once. They may ignore warning signs. They may have short fuses themselves. They may stick with counterproductive strategies and endure countless meltdowns in anticipation of some long-term gain. They may be stressed and frustrated about work, finances, schedules, and so forth. Their response to their inflexible-explosive child may be driven by the example they're trying to set for their other children.

In other words, there are many ways an adult may fuel a child's inflexibility-explosiveness by creating a distinctly user-*un*friendly environment. Many of these issues are discussed in greater detail in Chapter 8.

- **In a user-friendlier environment, adults try to use a more accurate common language to describe various aspects of the child's inflexibility-explosiveness.**

As you already know, I don't find a lot of the terminology commonly used to describe these children to be particularly useful, and neither do the children with whom I work. Tell a child he's bratty, manipulative, attention seeking, stubborn, controlling, and angry long enough, and an unfortunate thing may happen: He'll start to believe you. I find it more useful to talk to such children about being "easily frustrated"; how they have trouble "thinking clearly" when they're frustrated; things their parents can do to help them "stay calmer," "think things through," and "consider their options" when they're in the midst of frustration; the importance of recognizing the "early warning signs" of "vapor lock" and the circumstances that may lead to "meltdowns"; and the necessity of "basket thinking" and the "art of compromising." As you may imagine, children tend to be more receptive to being helped when this terminology is used. "Talking the talk" will make more sense as you progress through the next few chapters.

- **That vision thing.**

In some important ways, inflexible-explosive children are not what most parents hope for. Their difficulty adapting to curves in the road and extreme responses make life very difficult for them and those who must interact with them. They can affect family members and family life in potent, adverse ways. They can make parents question whether there's anything that's actually pleasurable and rewarding about being a parent. Yet, I've found that when we can start helping these children around the obstacles that are blocking their cognitive wheelchairs, they can be remarkably rewarding to have around. These are sweet, engaging, intelligent, interesting, curious, imaginative, creative, and sensitive children. But feeling good about them and helping them and you move beyond their difficulties require a change in vision. If you were hoping for the standard, compliant, easygoing child, it's not in the cards. One of the things I hope will happen as you continue reading this book is that you will form a more realistic vision of who your child is and what you can hope to create, both in your child and your relationship with him.

❄

Remember Marvin from Chapters 1 and 4? In one early session, I told his parents that once the user-friendlier environment kicked in I felt certain they'd discover they had a sweet son. I was confident in saying this because I'd seen clear signs of a great child underneath all the irritability, inflexibility, impulsivity, and explosiveness. The parents wanted to believe me, but they were skeptical. We saw a reduction in meltdowns and an improvement in parent-child interactions over the course of a few months. And, after trials of several types of medication, one finally started having the desired effect. Marvin became less moody and more malleable. He was less reactive and explosive. Little things didn't bother him as much. He became much more receptive to talking about and working on his difficulties.

Marvin's mother arrived for one of the next sessions in tears. I thought he'd had some sort of backslide.

"What happened?" I asked, expecting the worst.

"Oh, nothing . . . nothing bad anyway," she said in between sniffles. "It's just that you turned out to be right."

"Right about what?" I asked.

"Remember when you told us that underneath all his difficulties Marvin was really a very sweet kid? Well . . . you were right. He still has his moments, and we know things won't ever be easy. But we're finally enjoying him . . . we're seeing what a great kid he is . . . and that's something we weren't sure would ever happen."

Drama in Real Life
Shifting Gears, Exhibit A

April is a sixth-grade girl who managed to live the first eleven years of life without a diagnosis. Although she tended to be mildly impulsive, absentminded, and disorganized, she did well in school academically and behaviorally. Her behavior at home was a different story altogether. At home, April was prone to frequent, intense meltdowns during which she would scream and insist that her demands be met. Although she was neither physically aggressive nor prone to swearing during her meltdowns, her mother found April's inflexibility and low frustration tolerance difficult to deal with. The mother and daughter often ended up in shouting matches over even the most trivial issues.

In one early session, the mother recounted one of the memorable episodes of the week. "April was doing a report on Colombia in school and decided the night before she was to give the report that she wanted me to cook a traditional Colombian dish for her class. When I told her I didn't know how to cook any Colombian dishes, she went berserk on me. I finally started screaming back at her. Can you imagine?"

"Yes, I can imagine," I replied. "From what I gather, you and April have been screaming at each other for a long time now."

"I can't stand it," said the mother.

"Based on what you've described, it sounds like April is usually a very sweet, well-intentioned child," I said.

"Not a mean bone in her body," said the mother.

"And yet, there are times when she gets something in her head and can't seem to get it out," I said.

"Exactly," said the mother. "But what's the matter with her? Why does she start screaming over the smallest thing?"

"Probably because she can't think of anything else to do," I replied.

"She can't think of anything else to do?" the mother asked.

"That's right," I said. "She gets it in her head that she wants you to cook a Colombian dish for her class and then she's totally at a loss when you tell her you don't know how. She can't configure a way out of the dilemma. So she just keeps repeating the same request over and over."

"That's exactly what she does!" said the mother.

"The more she repeats the same request, the more aggravated you both become until you're finally screaming at each other," I said. "The problem is, telling her you can't cook a Colombian dish doesn't help her solve the problem."

"What should I do instead?" asked the mother.

"Help her think," I said. "When she's stuck like that—when she's showing you she can't shift gears—she's telling you she can't think of a way out of the situation on her own. She needs you to help her think things through. Your challenge is to make sure she stays coherent while you're trying to help her. Just remember, the closer to meltdown she gets, the harder it will be to help her think. The more aggravated you get—the more you raise your voice—the closer to meltdown she moves."

"Does this have a name?" the mother asked.

"I just call it inflexibility-explosiveness," I said. "I think that pretty much describes what's going on."

"You know," the mother said sheepishly, "I think I might be inflexible-explosive myself."

"Is that so?" I replied. "I'm glad you're sitting down because I have some very bad news for you."

"What's that?" asked the mother.

"I have more faith in your ability to help April be more flexible than I do in her ability to help you," I said.

"I had a feeling you were going to say that," said the mother.

"Luckily," I said, "I expect the help we provide to April to be very useful for you, too."

Drama in Real Life
Shifting Gears, Exhibit B

Jack (who you met briefly in Chapter 1) is an inflexible-explosive ten year old with Tourette's disorder and ADHD who has demonstrated an incredible capacity for melting down over just about anything since he was an infant. He lives with his mother and father, who, in striving to be good parents, were giving Jack plenty of opportunities to melt down. Jack's frequent, intense meltdowns—which included swearing and destroying property but no physical aggression directed at other people—were taking a major toll on his parents, who were at a loss for how to get Jack to go along more easily with their ambitious parenting agenda. Previous therapists had provided guidance on how to implement the standard behavior management approach, and the parents had given this approach their best shot on numerous occasions. But rewards seemed irrelevant to Jack when he was frustrated, and time-out and grounding eventually became impossible to enforce.

During the initial visit to my office, the parents complained about Jack's failure to respond to their demands, and Jack complained that his parents were always on his case. This is not an unusual scenario, but it's more typical of adolescents than of ten year olds. I asked the parents to try to come up with an itemized list of their parenting agenda—in other words, things they were constantly asking Jack to improve upon. The next week, they presented a list of over thirty goals, many of which were inducing daily meltdowns. I was struck by several things about the parents' list. First, most of the goals seemed perfectly reasonable. Second, there were way too many reasonable goals. Third, most of the goals weren't enforceable.

I tried to get a better sense of Jack's and his parents' capacity for flexibility.

"Jack, one of the items on your parents' list is for you

to spend less time with your friends so you're not so tired when it's time to do your homework," I said. "What do you think of that goal?"

"God, they never want me to spend any time with my friends!" Jack said, his voice rising.

"Jack, I don't think they want to deprive you of your friends; I think they just want you to spend a little less time with them," I clarified.

Jack's face started turning red. "That's not true!" he practically shouted. "They never want me to spend any time with my friends!" I was not exactly awed by Jack's capacity for flexibility so far. In fact, I was most impressed with how little frustration it took to send him into vapor lock.

Jack's mother could barely contain herself. She glared at Jack. "Well you can't spend as much time as you do with your friends, and that's the end of it!" Under some circumstances, I would have done something to prevent the impending meltdown at this point. But it was important for me to see what vapor lock, crossroads, and meltdowns looked like in this family.

"What are you dumb asses gonna do about it?" yelled Jack, turning red. "You can't control how much time I spend with my friends!"

"Well that's why we're here at the doctor's office," the mother glared. "And I refuse to be spoken to like that! The doctor is going to help us get you to act the way you're supposed to."

"I don't give a shit about the doctor or about you or about your stupid rules!" Jack screamed, storming out of my office.

His mother was seeing red. "Do you see what we have to deal with?" she demanded. "Do you see how he talks to us? I can't take this anymore!"

"It looks extremely unpleasant," I said. "I don't like the way Jack talks to you either. But I think there are a few things we need to think about. How much time do you actually want Jack to spend with his friends?"

"Well, he's with them all afternoon on school days and most of Saturday and Sunday. We think that's just too much," said the mother.

"At the moment, do you have any way to force Jack to spend less time with his friends?"

"That's what we were hoping you could help us with," interjected the father.

"How have you tried to enforce your wishes already?" I asked.

"You name it," said the mother. "Sticker charts, rewards, time-outs, grounding . . . he's climbed out his window when he was grounded several times. After all that, he still spends too much time with his friends. Then we pay the price when he goes crazy because he doesn't have enough time to get his homework done."

I stated the obvious. "So I gather there really is no way to enforce your wishes. And the only thing that happens when you raise this issue with Jack is more anger and hostility."

"So what are we supposed to do, let him spend all his time with his friends?" the mother seethed.

"Well, I guess I don't expect to make a lot of progress insisting on any of your goals if we don't have ways of enforcing them," I said.

"Any ideas?" the father asked.

"To be perfectly honest, I'm not eager to have you try things that haven't been effective before," I said.

"So now what?" asked the mother.

"Well," I said, "if most of your goals aren't enforceable, you and Jack will have to do a better job of talking and negotiating with each other."

"Negotiating?" boomed the mother. "These goals are nonnegotiable!" she insisted.

My assessment was that neither the mother nor the son were endowed with an overflowing capacity for flexibility, and I wasn't sure about the father yet because he hadn't said much. There were also some hints that the mother wasn't very good at recognizing vapor lock in her son—perhaps because her own frustration kicked in so fast—and that she ended up fueling his progression toward full-fledged meltdowns.

In our next session, Jack's parents recounted his most memorable meltdown of the week. It seemed that on Friday night, Jack called his mother from a friend's house to inquire about the evening's activities. Jack's mother told him that she and his father were going to the movies and invited Jack to come. Jack immediately asked if he could bring his friend Louis.

"No way, Jack, you've got the whole weekend to be with Louis," the mother responded tersely.

"What's the big deal about me bringing Louis to the movies?" Jack demanded (vapor lock).

"I don't feel like being a chauffeur for you and your friends again tonight, Jack," the mother replied. "And I don't feel like arguing about it."

"How about you go to your movie and drop me and

Louis off at a different movie?" offered Jack, demonstrating some capacity for rational negotiating (crossroads).

"I'm not taking Louis to the movies tonight, Jack," the mother said. "What is it about that statement you don't understand?"

"What's the big deal?" Jack screamed into the phone.

"Bye, Jack," said the mother, hanging up.

Jack arrived home minutes later on the verge of a total meltdown.

"You hung up on me!" he shouted at his mother. "Why can't Louis come to the movie with us! What's the big fucking deal?"

"Jack, I told you, I don't feel like having one of your friends tagging along tonight, and that's the end of it," said the mother, in one last valiant attempt to remind Jack who was really the boss.

It wasn't the end of it. Indeed, it was only the beginning. Enter Jack's father.

"What's the big deal about him taking a friend?" he asked the mother. "Is it worth all this screaming?"

"What's the big deal about me not wanting him to?" yelled the mother. "I can't deal with this tonight!" The mother angrily stalked upstairs and slammed the door to her bedroom.

"Well, if I can't go to the movies, then I want to sleep over at Louis's house," Jack demanded.

"You can sleep over Louis's house tomorrow night, but not tonight," said the father.

Commence meltdown.

"You assholes said I can have one sleepover per weekend, and I want to sleep over Louis's house tonight!" Jack yelled.

"I don't want you to sleep over Louis's house tonight, and I don't like being spoken to that way," said the father.

"Jesus, you guys are such fucking liars!" Jack yelled, his frustration at a peak.

"I don't like being spoken to that way," the father repeated, "and if you continue, you'll be grounded for the weekend."

"Screw you!" Jack screamed.

Jack was then told he was grounded. On his way up to his room, he kicked a hole in the wall. The meltdown lasted a full two hours.

After the parents finished telling the story, I carefully asked the following question: "How important was it to you that Jack's friend not go to the movies with you?"

Jack's mother thought for a long moment. "In retrospect," she replied, "not very."

I then asked the father a similar question: "How important was it to you that Jack sleep at his friend's house Saturday night instead of Friday night?"

The father replied, "Not very." Both parents were then asked whether either issue merited a two-hour meltdown. "No way," was their collective response.

"It seems to me," I said, "that if an issue isn't very important—and not worth enduring a meltdown over—and Jack doesn't seem to be learning anything from the meltdown anyway . . . then these may not be issues worth inducing and enduring meltdowns over."

"You don't think Jack learns anything from these episodes?" the father asked.

"I don't know . . . let's see," I said, turning to Jack. "What did you learn from this episode?" I asked.

Jack responded, "God, my father can't even take a little insult. And what was the big deal about me taking a friend to the movies?"

"Do we have any evidence that Jack actually learned anything from this episode?" I asked the parents.

"Evidently not," replied the father.

"Do we have any evidence that Jack has ever learned *anything* from past meltdowns? In other words, have all these meltdowns actually made it less likely that Jack will explode the next time there is a disagreement?" I asked.

The parents reflected briefly. "No, probably not," replied the mother.

"So why are we inducing and enduring so many meltdowns?" I asked.

The mother responded defensively. "Because we're the parents and he's the kid."

"Well, I think that there are some issues that are worth inducing and enduring meltdowns over—like safety—but it looks like Jack is doing whatever he wants to do right now anyway," I said.

"Should we have just let him take a friend to the movies?" asked the father.

"At this stage of the game, I think so," I replied. "I'm not convinced that preventing Jack from taking friends to the movies is real high on your list of priorities right now. In fact, it seems to me that reducing meltdowns is far and away your highest priority. So when faced with the choice of either letting Jack take his friend to the movies or inducing and enduring a meltdown, it seems that the higher priority is to avoid the meltdown. If we reduce the number of things Jack is melting down over, we'll be able slowly to reduce the overall tension in your household. Then we need to help you and your son start talking to each other. Then maybe we'll be able to start talking about some of these goals you're interested in pursuing."

"Whoa, you want us to ask *less* of him?" asked the mother.

"Well, yes," I replied. "I don't expect things to improve by my helping you ask *more* of him, since you're getting very little of what you want right now. And I don't have any reason to believe that the motivational strategies that didn't work very well for you before will work any better for you now. So let's start talking about what I think *will* work."

❄

Before we move on to basket thinking, a brief summary of this chapter may be useful.

- The approach described in this book is comprised of two basic avenues. The first avenue involves making various adjustments so as to make the environment user-friendlier for your child. Creating a user-friendlier environment can clear the smoke and help set the stage for you and your child to begin working together to deal more effectively with his inflexibility-explosiveness.

- First and foremost, a user-friendlier environment should help you respond to your child before he's at his worst, rather than when or after he's at his worst. Here are the specific components:

1 Making sure all the adults who interact with your child have a clear understanding of his unique difficulties, including the specific factors that fuel his inflexibility-explosiveness.

2 Reducing the overall demands for flexibility and frustration tolerance that are being placed on your child, by judiciously establishing priorities among your parenting goals.

3 Identifying—*in advance*—specific situations that may routinely lead to inflexible-explosive episodes.

4 Reading the warning signals and taking quick action when these signals are present.

5 Interpreting incoherent behaviors for what they really are: incoherent behaviors.

6 Understanding the manner in which you and other adults may be fueling your child's inflexibility-explosiveness.

7 Using a more accurate common language to describe various aspects of your child's inflexibility-explosiveness.

8 Coming to a more realistic vision of who your child is and what you can hope to create, both in your child and in your relationship with your child.

7
Basket Case

One of the most important aspects of my work with inflexible-explosive children and their families is repetition. Old habits die hard and old communication patterns are hard to break, so repetition is a crucial ingredient for keeping things moving in the right direction. Toward this end, here are some of the critical points we want to keep fresh:

- Flexibility and a tolerance for frustration are skills. Because of any variety of factors, your child is predisposed to have difficulty responding to the world in an adaptable, flexible manner.

- Because inept parenting, poor motivation, attention seeking, and the lack of appreciation for who's boss may not be the best explanations for your child's difficulties, standard parenting practices and motivational programs may be mismatched to his needs. Your child may require a different approach.

- How you perceive and understand your child's inflexibility-explosiveness is directly related to how you'll ultimately respond to it. If you respond to your child in an inflexible, angry manner, then you will increase the likelihood of meltdowns, fuel adversarial parent-child interactions, and reduce the likelihood of improving your child's flexibility.

- As you've probably already discovered, it's unlikely that the meltdowns you've been inducing and enduring have taught your child anything productive or led to any meaningful positive change in his behavior.

- I am more optimistic about your ability to respond differently to your child's inflexibility than I am about your child's ability radically to improve his capacity to be flexible, at least initially.

- If you create a user-friendlier environment and concentrate more on the front end (before and during vapor lock and crossroads) than the back end (during and after meltdowns), interactions between you and your child should improve. Once things begin to improve a little, he should become more receptive to, and have a greater capacity to benefit from, more direct methods for increasing his flexibility and frustration tolerance.

The last chapter reviewed important components of a user-friendlier environment, including adjusting your parenting priorities to reduce the overall demands for flexibility and frustration tolerance being placed on your child, identifying specific situations that may routinely lead to inflexible-explosive episodes, reading the warning signals of vapor lock so you can take quick action before your child melts down completely, and accurately interpreting the mental debris flowing out of your child's mouth during meltdowns. You also read about some strategies that may be helpful in the early phases of vapor lock. Let's now turn to a simple framework for helping you incorporate many of these components.

This framework is built around two basic themes: **First, *reducing* the frequency of your child's meltdowns has to become your top parenting priority. Second, helping your child *maintain* coherence in the midst of frustration has to become your main goal during vapor lock; *restoring* coherence has to become your main goal during meltdowns.**

Let's think about the first theme briefly. Different parents have different priorities for what they try to teach their children, and these priorities tend to influence parents' decisions about whether, when, and how to intervene in given situations or with respect to certain behaviors. For example, my sister's top priority for her children is that they be very independent. My sister-in-law's top priority for her children is that they be very clean. So, imagine my sister's two children and my sister-in-law's two children playing in a mud puddle. My sister would likely glow with pride at the sight of her two mud-covered children, pleased that they had thought of the idea on their own and exhibited leadership qualities in convincing their two cousins to join them in muddy pursuits. It's unlikely she'd feel the need to intervene, except perhaps to shower her children with praise. My sister-in-law, in contrast, would take one brief glance at her soiled sons and pass out. When she came to, she would definitely intervene, probably by showering her children with soap and water. The same scenario, but different priorities and different reactions.

Parents of inflexible-explosive children don't really have the luxury of pondering their top priorities. Because of the incredibly destructive impact of meltdowns, reducing the frequency and intensity of meltdowns has to become their number-one priority. So while parents of children who are not inflexible and explosive can be more relaxed and reflective about their priorities, parents of children who are inflexible and explosive must walk a narrower path because the stakes are much higher.

Thus, when parents of inflexible-explosive children see their child doing something inappropriate or have a task they want their child to perform, they have to ask themselves the following question: **"Is this behavior important or undesirable enough for me to induce and endure a meltdown?"**

In other words, the parents must constantly decide whether their child's compliance with a given expectation is worth the price of a major, destructive, labor-intensive meltdown. Is a messy bedroom important or undesirable enough to induce and endure a meltdown? An unmade bed? Unbrushed teeth? Incomplete homework? Bad manners? Eating unhealthy foods? Destruction of property? Swearing?

If you want to restore some level of sanity to your household and start helping your child be more flexible and less explosive, then for now—and maybe for quite some time—reducing the frequency of your child's meltdowns will have to rank higher on your list of parenting priorities than a lot of other priorities. That being the case, your answer in many situations will be, "No, this behavior is not important or undesirable enough for me to induce and endure a meltdown." In so responding, you'll purposely choose to avoid a meltdown and consciously decide that family peace, improving your relationship with your child, and setting the stage for helping your child be more flexible and handle frustration more adaptively are higher priorities than a bed being made or a room being cleaned or homework being done. As you already know, your other option is to continue routinely inducing and enduring meltdowns in the hope that such meltdowns will lead to a productive outcome. As you also know, my sense is that if inducing meltdowns hasn't been a useful teaching tool for the past two to ten years, inducing several hundred more meltdowns for another year probably won't be either. This theme definitely takes some getting used to. Don't stop reading yet.

If you want to restore some level of sanity to your household and start helping your child be more flexible and less explosive, then for now—and maybe for quite some time—reducing the frequency of your child's meltdowns will have to rank higher on your list of parenting priorities than a lot of other priorities.

Think about the second theme for a moment. Because your top priority is to keep your child from melting down, it is in your best interests to help your child maintain coherence and rationality during vapor lock. The more incoherent your child becomes in your attempts to get him to shift from Agenda A to Agenda B, the closer to meltdown he'll move. And the closer to meltdown he moves, the less likely he'll be to be able to shift from Agenda A to Agenda B. (It follows that the more incoherent *you* become as you attempt to get him to shift from Agenda A to Agenda B, the closer to meltdown he'll move.) Anything you do to fuel your child's cognitive and emotional overload during vapor lock—insisting on his rapid compliance on low-priority issues, becoming angry, threatening, screaming, or berating—decreases coherence in your child and increases the likelihood of a total meltdown. Even your attempts to reason with your child may feel overwhelming to him; if his capacity for coherent thought is already depleted, your attempts to engage in logical discourse may end up adding to his frustration. Your "reading" skills will prove quite important along these lines.

Of course, it's not so easy to sort through all this when your child first enters vapor lock. In other words, it would help if you had a simple way to decide how to answer the question, "Is this behavior important or undesirable enough to induce and endure a meltdown?"

Picture three baskets in a row: Basket A, Basket B, and Basket C. In Basket A are the behaviors that *are* important or undesirable enough to induce and endure a meltdown. In other words, Basket A contains those behavioral expectations that are nonnegotiable. Basket B is the "wheelchair-ramp" basket. It contains behaviors that are important or undesirable but over which you're not willing to induce and endure a meltdown. Basket B behaviors are negotiable. And in Basket C are behaviors that aren't important or undesirable enough even to say anything about anymore. In other words, these behaviors are off the radar screen. Let's take a closer look inside each of these baskets.

Basket A

As you just read, the behaviors that are in Basket A are nonnegotiable. Basket A doesn't have many behaviors in it. Because your number-one goal is to reduce the frequency and intensity of your child's meltdowns, there aren't going to be many behaviors that are important or undesirable enough to induce and endure meltdowns over.

Safety is always in Basket A. In other words, unsafe behaviors—defined as those that could be harmful to your child, other people, animals, or property—are not negotiable and are worth inducing and enduring meltdowns over. Of course, just because you're willing to induce and endure a meltdown to ensure safety doesn't mean a meltdown is your goal. If your child engages in an unsafe behavior—hitting, kicking, throwing things, destroying property, and the like—you're taking the least intense, least physical, least antagonistic route to restoring safety and coherence. But you're not negotiating.

For many inflexible-explosive children, unsafe behavior is the only thing placed in Basket A early on. These children may be exploding so often and so violently that safety is about all they can work on initially. The good news is that through the creation of a user-friendlier environment (and sometimes medication), the majority of inflexible-explosive children are able to stay safe even when they become frustrated, and this alone can have a dramatic, positive effect on family interactions. Indeed, it's hard for parents and siblings to remain invested in the user-friendlier theme as long as unsafe behavior persists.

To keep Basket A fairly empty—and avoid some of the parenting pratfalls that could undermine the user-friendlier environment we're trying to create—the following litmus test should be applied to any behaviors being considered: Not only must the behaviors be important or undesirable enough to induce and endure meltdowns over, they must also be behaviors that your child is capable of successfully exhibiting on a fairly consistent basis and those that you're actually willing and able to enforce. There's no sense inducing and enduring meltdowns over expectations your child is incapable of meeting or behavioral limitations you aren't willing or able to enforce.

As you've read, early in treatment I often ask the parents of an inflexible-explosive child to give me a written list of the behavioral priorities their child is not meeting on a consistent basis. It's not unusual for me to be greeted with a long list. I typically inform the parents that their priorities are going to need some serious pruning. If you overburden your child with a lengthy list of priorities—even if they're all ones you're willing and able to enforce—you'll actually increase the number of opportunities for meltdowns, but you most assuredly won't increase the likelihood that your priorities will be met. It follows that if you reduce the priorities to a number that is more realistic, given your child's actual capacities—by temporarily suspending your desire for, and apparent willingness to induce meltdowns over, things like good manners, clean rooms, completed homework, and so forth—you'll simultaneously reduce the opportunities for vapor lock and meltdowns, lower the general level of family tension, and quite possibly get better compliance on the demands you actually do keep in Basket A. Once the number of meltdowns begins to decline—and not a moment before—you may be able to begin slowly expanding your list of undesirable and important behaviors again.

The truth is, even parents of flexible children can't intervene every single time they see their child engaged in an undesirable behavior; they understand that they have to let some behaviors slide without correction, fully appreciative of the old "pick your battles" adage. This formula applies to your child as well—just more so.

The parents of a child I was working with had decided they wanted him to reduce his sugar intake because they thought his consumption of soda and candy adversely affected his mood and behavior. This goal would have been fine, I suppose, except for two significant problems: The child had no interest in reducing his sugar intake and, worse yet, he was consuming most of his sugar outside the home, making this priority virtually impossible to enforce. Unfortunately, the parents had skipped right past the Basket A litmus test and had been jumping on the child's case for alleged sugar consumption whenever he came home from being with his friends. This practice induced meltdowns with predictable regularity. I was able to convince the parents that since they could not enforce this priority, sugar consumption should be placed in either Basket B or C, but absolutely not in Basket A. Their son ended up consuming no more or less sugar as a result of their moving this priority to a different basket, but he melted down about a dozen fewer times each week.

So what else may end up in Basket A besides safety? Some inflexible-explosive children refuse to go to school, have great difficulty getting there on time, or skip out once they do manage to get there. These school-related issues often find their way into Basket A. The trouble is, these problems don't typically occur out of the blue. In other words, when children feel that school is a place where they can experience success, they are generally willing to go there, get there on time, and stay put. It's when things aren't going so well—either at school or in other areas of children's lives—that these problems tend to arise. So I sometimes hold off on placing these problems in Basket A until I've had a chance to assess and address what's not going so well.

Difficulty getting going in the morning and settling down in the evening are common in children with ADHD and mood disorders. Lots of inflexible-explosive children are still revved up at 9 P.M. when you wish they'd start fading. At 10 P.M., some are just beginning to work on the homework you couldn't force them to do earlier in the day. Others are so overtired at night that their moods have begun to deteriorate. The next morning they're dead to the world or, at best, extremely agitated the minute they manage to get their eyes open.

Sometimes medication can resolve these difficulties, sometimes not. Frequently, the best we can do nonmedically in the morning is make sure that demands are sharply reduced, so the child needs only to get dressed (sometimes with help) and get out the door (perhaps with a banana firmly planted in his hand on his way out). There are lots of perfectly reasonable priorities that may present unnecessary roadblocks to a child whose brain hasn't clicked on yet in the morning, such as eating breakfast, brushing his teeth, combing his hair, taking out the trash, walking the dog, and packing the backpack. Many of these priorities ultimately end up in Basket B or C, which means dropping the priority completely, permitting the child to delay its execution until later (for instance, taking a shower after school), or performing certain tasks for him (he may never learn to pack his backpack himself, but that's probably not going to keep him from getting into Harvard). The best we can usually do nonmedically at night is enforce getting the child to stay in his bedroom at a designated time.

My view on homework is a little controversial because, at least initially, I'm not enthusiastic about putting the completion of homework into Basket A. In other words, I often find that I can make a major dent in the frequency and intensity of meltdowns by throwing homework out of Basket A. Often I'll try to get the child's school to take on the responsibility of overseeing his completion of homework, especially if the child isn't prone to melt down at school. Frequently, especially in the case of children with ADHD who are unmedicated during homework time, I help the parents and teachers put their heads together so homework assignments are judiciously prioritized and homework time is reduced to the barest necessities. In the case of severely explosive children, sometimes we simply decide that the child won't be doing much homework for a while until we can get the explosions under control.

Swearing is not a safety issue, as excruciating as it is to listen to. Once we give these children a new language to express their frustration, the swearing usually begins to subside. I often ask adults what *they* do when they become frustrated and have found swearing to be a remarkably popular response—which begs the question of why we would expect children to respond more adaptively to frustration than we do ourselves. I say more about swearing in later chapters.

If you've been bothered by the nagging feeling that averting meltdowns means you'll have little authority over your child, Basket A should actually comfort you because it signifies that there are some things your child has to do just because you've said so. Just not very many. The number of behaviors in Basket A increases as your child's capacity to deal with more Basket A behaviors increases. But Basket A isn't the most important basket. Most inflexible-explosive children are already convinced of your status as an authority figure; the problem is that when they're incoherent, they have trouble acting on this knowledge.

Basket B

The most important basket, in fact, is Basket B, which includes behaviors that you've decided are high priorities but over which you are not willing to induce a meltdown. What do you do instead? Find a wheelchair ramp. Help him think. Communicate. Negotiate. Compromise. The dictionary defines *compromise* as **a settlement in which each side gives up some demands or makes concessions; something midway between two other things.**

Most inflexible-explosive children are already convinced of your status as an authority figure; the problem is that when they're incoherent, they have trouble acting on this knowledge.

Compromise is, for a lot of inflexible-explosive children, the pathway to hanging in and staying coherent in the midst of frustration, so they can resolve conflict in a mutually satisfactory manner. Compromise is your way of getting *some* of what you want from your child without inducing a meltdown.

Many inflexible-explosive children have atrocious negotiating skills. You may be thinking, "Wait a second . . . my child negotiates with me all the time." Many parents who tell me this are really saying that their child is incredibly persistent but not particularly reasonable in his persistence. Often it turns out that the child is actually quite limited in his ability to engage in the give-and-take required to reach solutions that are mutually satisfactory. So we've got a lot of teaching and modeling to do. Compromising is hard for many parents, too; although most figured they'd eventually have to start negotiating with their child once he reached adolescence, they never envisioned the necessity of negotiating with a six year old. (Of course, there are many things parents never envisioned before they discovered they had an inflexible-explosive child.) So compromise is one of the most important skills we're going to try to teach. Instead of training rapid compliance, you'll be training a skill that may be much more important.

Here's how it often sounds when we're trying to teach a child about the art of compromise, with appropriate modifications depending on the child's age:

Me: Johnny, do you know what a compromise is?

Johnny [*age seven*]: Sorta.

Me: What's a compromise?

Johnny [*brief pause*]: It's when everybody gives in a little.

Me: Well, that's pretty close . . . it's when everybody's a little happy because they got some of what they wanted, but nobody gets everything they wanted.

Johnny: Yeah.

Me: So, should we try one?

Johnny: One what?

Me: One compromise. You know, to see if you can do it.

Johnny: OK.

Me: If your mom wanted to go skating at four o'clock and you wanted to go to skating at two o'clock, what would a good compromise be? [Don't forget, Johnny's not very good at thinking about compromise yet, so I don't expect much from him at this point.]

Johnny: Uhm . . . I dunno . . . we could skate at two o'clock *and* four o'clock.

Me: That's an idea, but don't forget, your mom doesn't want to go skating at two o'clock. Can you think of a way to make her a little happy and you a little happy?

Johnny [*brief pause*]: Nope.

Me: Well, what's in the middle of two and four o'clock?

Johnny: Three o'clock.

Me: So do you think three o'clock would make you a little happy and your mom a little happy?

The child often looks to his parents for approval of this high-stakes deal.

"Would it?" he might ask. The parents, surprised that the child is even participating in this discussion, vigorously nod their approval.

Me: I think that would be a good compromise. OK, let's try a harder one. Let's say you wanted to go sailing today, and your friend wanted to go bowling. What would a good compromise be?

Johnny: We could go bowling today and sailing

tomorrow, or we could go sailing in the morning and bowling in the afternoon. Right?

Me: Those would be great compromises, Johnny. I think you've got the hang of it. This week when you and your mom or dad disagree about something, they'll let you know whether they're willing to compromise. If they are, try to think of good compromises. Just remember, your parents have the final say on any compromise, so try to come up with compromises that you think your parents will like. If your first idea doesn't work, keep coming up with ideas until you and your mom or dad find a compromise you all can agree on.

Because inflexible-explosive children have difficulty thinking rationally during vapor lock and usually require some practice thinking about how to compromise, getting good at generating compromises may take longer than you'd think and may require additional practice at calmer times. Some have a hard time generating compromises on their own initially. What this means is that, early on, you may have to provide suggestions for a lot of the compromises. It also means that, early on, you're probably best off practicing compromise on issues on which you're likely to succeed in reaching a mutually satisfactory agreement.

It's also helpful to the child for you to summarize the disagreement that needs to be compromised on. In other words, if the child becomes disorganized in the midst of frustration, he may need help figuring out what the disagreement is all about before he can actually try to generate possibilities for compromise. The summary would sound something like this:

"Johnny, you want to go sailing today, and your brother wants to go bowling. So there's a disagreement between you two. We need to think of a good compromise."

As this dialogue suggests, it's often appropriate to bring siblings into the compromising mix. The siblings will need to be trained how to compromise, too, of course, but they'll appreciate the eventual absence of screaming and the fact that their needs are being taken into account.

In addition to helping children think more clearly, compromising can also help them remain coherent in the midst of frustration. They'll need a lot of help here as well. It might sound something like this:

Parent: Sally, it's time to come inside.

Sally [*vapor lock*]: I want to stay outside and play a little longer.

Parent: Uhm . . .

Sally [*deeper vapor lock*]: You always make me come in before everyone else! It's not fair! I hate . . .

Parent: Whoa! Slow down! I don't really think this is one to get all upset about. I may actually be willing to compromise about this if you could chill a little.

Sally [*more coherent*]: Everyone else gets to stay out longer than me.

Parent: Yes, so I heard you say. Let's think about this for

a second. I want you to come in right now and you want to stay out longer. How much longer do you want to stay out?

Sally: Twenty minutes!

Parent: Hmm. Sounds like we definitely need to compromise. You want to stay out twenty more minutes and I was thinking more like ten extra minutes. Can you think of a good compromise?

Sally: No! Ten minutes isn't enough!

Parent: Now, don't go getting yourself all worked up again. . . . I told you I'd be willing to compromise. But we have to figure this out. Let's try to think of a good compromise.

Sally: I can't do this! I can't think!

Parent: Well, fifteen minutes is in between ten and twenty. . . . would fifteen minutes be a good compromise?

Sally: Okay, yes! Fine!

Parent: Before you go running off, I just want to remind you that in fifteen minutes I expect you to come in with no problem at all. Do we have a compromise?

Sally: Yes! Geez!

What if it begins to appear unlikely that a compromise will be reached on a Basket B issue? In some instances, that's when a Basket B issue ends up defaulting into Basket C. Remember, if you've decided that a behavior doesn't belong in Basket A, you've already decided that the behavior is not worth a meltdown. Just because you can't reach a compromise doesn't mean you're now willing to induce and endure a meltdown. If you can't reach a compromise, the issue drops into Basket C, and you move on.

Here's how such a scenario might go:

Greg: I want to sleep over Andy's house tonight.

Parent: Greg, you've got a lot of homework to do this weekend, and you've been very tired lately. I want you to stay home tonight.

Greg [*entering early vapor lock*]: No way! It's a weekend night, and you always let me sleep over someone's house on the weekend. Come on!

Parent [*recognizing early vapor lock, quickly deciding that this is not a Basket A issue, and maintaining a neutral tone*]: I think we need to find a compromise here. I want to make sure you get your rest, and you want to sleep over Andy's house tonight.

Greg [*progressing further into vapor lock*]: I always get my homework done, and I'm not tired! You always do this to me!

Parent [*now recognizing the potential for meltdown, cog-*

nizant of the possible need to dump this issue into Basket C, and still maintaining a neutral, matter-of-fact tone]: I'm not saying you can't sleep over Andy's house tonight, I'd just like us to talk a little about how we're going to make sure you're not bushed when you come back home tomorrow.

Greg [*becoming exasperated and now teetering on the edge of the "coherence cliff"*]: I can't think of a compromise! I think compromise is stupid!

Parent [*not becoming exasperated and not teetering on the edge of a meltdown*]: Well, one possible compromise is for you to promise me that you'll make sure you and Andy are asleep by midnight.

Greg [*still teetering but hanging in*]: You know I can't do that! Why are you doing this to me!

Parent: Another possibility is to sleep over Andy's house tomorrow night.

Greg [*teetering but still hanging in*]: He's busy tomorrow night!

Parent [*having now decided that compromise is unlikely under the circumstances and having therefore determined that this issue has defaulted into Basket C*]: Well, why don't you go ahead and sleep over Andy's house tonight and try your best to get a decent night's sleep.

Greg: Fine!

And then they let it go.

I've met many parents who have trouble sorting things through themselves when their child enters vapor lock. Fortunately, some children will tolerate the uncertainty of a delayed basket decision for a few minutes while their parent sorts through his or her own thoughts. In other words, it's not always critical that a basket placement be made immediately. If you think you can maintain your child's coherence while you buy time, you can say something like, "I haven't decided yet whether that's negotiable or not—you'll have to wait a few minutes so I can decide." Some children can handle a brief delay, others a longer delay, some no delay at all. And some of those who can't tolerate a delay eventually can once the user-friendlier environment takes hold.

Other parents feel the need to delay a basket decision because they'd prefer not to have such a dialogue in the car, supermarket, or shopping mall (places where their capacity to deal with their child's worst—if it happens—is compromised). When vapor lock sets in, these parents do a quick "probability estimate" of the likelihood they'll be able to deal successfully with a given issue under their current circumstances. If they aren't optimistic, they try to delay.

> A mother came in for a session one night hoarse from all the screaming she'd been doing at her daughter on the drive to my office.
>
> "What were you screaming about?" I asked.
>
> "Alycia was very upset that we're going to have to change our plans for her birthday," she replied.
>
> "That must have been very upsetting to her. But why all the screaming?" I asked.

"Because it was foggy out, and I'm not the best at driving at night to begin with, and I've got a frustrated daughter sobbing in the backseat telling me I don't love her," she replied.

"What did you do?" I asked.

"I screamed at her," said the mother. "And now I'm sitting here really mad at myself for doing it. I guess I get a little worked up when she gets worked up."

"What was your goal when she started getting upset?" I asked.

"I have no idea," replied the mother. "I just wanted to get past the problem, be done with it."

"That's an interesting goal," I said. "Because you have a daughter who's not very good at just getting past problems and being done with them."

"That's for sure," the mother agreed. "So what should my goal be?"

"I think one important goal is to think about whether the ideal time and place to have a discussion that's going to be frustrating to Alycia is in the car at night when it's foggy. In other words, whether it's at all likely a productive discussion could take place under those conditions. If you decide that scenario probably wouldn't set the stage for the discussion you want to have with her, you could try to delay the discussion until the circumstances are more ideal. Do you think Alycia would have been able to delay the discussion?"

"She still would have cried in the backseat," the mother said.

"Well, we can't keep her from feeling what she legitimately feels about her birthday plans being changed. But we can give some more thought to whether delaying the discussion has the potential to lead to a more productive outcome."

The truth is, delay can work in the child's best interests, too, since some inflexible-explosive children are able to think more clearly if the Basket B discussion is delayed.

It's in Basket B where you're helping your child develop skills that he may be sorely lacking, including hanging in, in the midst of frustration; taking another's perspective; generating alternative solutions; and thinking things through. It's also in Basket B where your child recognizes that you're able to help him learn these things.

It may be obvious that there's a lot more to Basket B than just averting meltdowns on important issues. It's in Basket B where, over time, you're helping your child develop skills that he may be sorely lacking, including hanging in in the midst of frustration; generating alternative solutions; and thinking things through. It's also in Basket B where your child recognizes that you're able to help him learn these things.

Basket C

Basket C consists of those goals and behaviors that may once have seemed a high priority but have since been downgraded considerably. Basket C is very full—that is, full of behaviors you're going to forget about completely, at least for now. Basket C is important because it contains many behaviors that will no longer frustrate your child. So Basket C helps us work toward the goal of reducing your child's global level of frustration (lowering the temperature on his frustratometer), which should pay dividends the next time he does become frustrated.

The idea is that if a behavior is in Basket C, you don't even mention it anymore, let alone induce meltdowns over it. Remember the boy who loved sugar? A few weeks after his mother and father had concluded that sugar consumption probably belonged in Basket C, he and his parents were in my office having a discussion. At some point during the discussion, his mother became agitated over something and headed straight for the topic of her son's sugar intake. "He's still eating too much sugar," the mother seethed at her son. She then swung around in her chair and seethed at me. "But we're not allowed to talk about that anymore." This, of course, was a no-no, for it defeated the whole purpose of Basket C.

Which behaviors should be in Basket C? A lot of them, but the specifics depend on you and your child. A few more stories may be instructive. One mother thought it was imperative for her inflexible-explosive daughter to wear gloves while she was snow skiing. Each week the mother and daughter happily arrived at the ski lodge. Each week the child melted down when the mother wanted her to put gloves on. Each week, after thirty to forty-five minutes of lunacy, the mother would carry her daughter, kicking and screaming, out to the car and drive home.

"Why doesn't she want to wear the gloves?" I asked in one early session.

"She hates the way they feel," the mother replied.

"Will she wear them if her hands get cold?" I asked.

"Yes, but I want her to wear them all the time," replied the mother.

"You want her to wear the gloves so badly that you're willing to endure a thirty- to forty-five-minute meltdown and a kid who kicks you and screams at you and embarrasses you and ends up not doing any skiing?"

"Well, I think it's important that she wear gloves, don't you?" the mother asked.

Actually, no. The gloves belonged in Basket C. It took awhile to convince the mother of this logic, but the daughter wound up doing a lot more skiing and a lot less melting down, and she never did let her hands get frostbitten.

Another child was remarkably particular about what foods he was willing to eat: certain cereals for breakfast and pizza for dinner. His parents were quite determined—manifested in their relentless badgering—that he have a balanced diet, but weren't willing or able to force lima beans down their son's throat. This example of reciprocal inflexibility led to at least two meltdowns a day, at breakfast and dinner. Except in extreme cases, such as bona fide eating disorders, issues associated with diabetes, and so forth, I have what can best be described as a Basket C mentality toward these picky-eating inflexible-explosive children: *They won't starve*. And, indeed, this child wasn't starving. We placed "eating a variety of foods" in Basket C, reduced meltdowns, improved communication, went after our higher priorities, and eventually were able to put food variety into Basket B. The child is now eating a somewhat wider range of foods, and he actually goes to the supermarket with his mother to make his own selections.

Who decides what basket a behavior falls into? You. Who decides whether an acceptable compromise has been reached? You. Who decides whether a child is actually capable of following through on a suggested compromise? You. Who's still in charge? You.

❄

Now you've got a basic, user-friendly framework that helps *you* decide what *you* want to happen before or when your child enters vapor lock. Basket A helps you ensure safety and maintain your role as an authority figure. Basket B helps your child maintain coherence in the midst of frustration; think through mutually satisfactory solutions; and view you as a helper, rather than as an enemy. Basket C helps remove many unnecessary frustrations from the path of your child's cognitive wheelchair. From now on, before or when your child enters vapor lock, your first step is to ask yourself, "Is this behavior important or undesirable enough for me to induce and endure a meltdown?" If the answer is no, your next step is to determine whether the behavior is even a high priority ("Is the behavior in Basket B?"). If so, you'll embark on a process of trying to negotiate a compromise. If not, the behavior falls into Basket C, and you let it go.

Many parents are skeptical when this framework is first described, so let's make sure you're crystal clear about what we're trying to accomplish here. Your frustration over your child's difficulty adhering to your parenting agenda, combined with (1) your child's own frustration over his difficulties adhering to this agenda and (2) his natural predisposition to handle frustration poorly, has led to (3) communication patterns that can best be described as unhelpful and (4) a child who can best be described as inflexible-explosive. To deal effectively with this set of circumstances, we need to prioritize a reduction in meltdowns, temporarily suspend much of your parenting agenda, reduce frustration all the way around, and work toward a more adaptive pattern of communication to set the stage for the development of skills that the child lacks. This won't happen overnight, and it won't be easy. But you've been working very hard already. Let's just make sure your hard work is getting you somewhere.

"Giving in" is now part of your old language for describing interactions between you and your child. What you're actually doing is judiciously choosing when, where, and especially over what issues he has a meltdown. In so doing, you're exerting more control—not less—over his behavior than before. In other words, starting now, it is you—not your child—who is the primary determinant of whether or not he has a meltdown. So, instead of feeling like you have no control over your child, you'll start finding that you actually have a great deal of control. By replacing your fear of meltdowns with a sense of empowerment over when and under what circumstances meltdowns occur, you're putting yourself in a much better mind-set for responding to vapor lock calmly and rationally, thereby helping your child respond to vapor lock calmly and rationally.

Remember, your best defense against meltdowns is to make sure they don't happen. What should you do if your child has a meltdown anyway, either over a Basket A issue or because you couldn't find a wheelchair ramp quickly enough? Do whatever it takes—distraction, empathy, comforting, separating, suggesting an alternative activity—to defuse the situation, restore coherence, and ensure safety as quickly as possible. In their calmer moments, children are often able to suggest the best ways for you to help restore coherence ("Leave me alone and I'll calm down" is a common response). You've already learned that being highly punitive during your child's meltdowns doesn't enhance his learning. And getting really angry at your child doesn't help him regain coherence any faster. Unless safety is at stake, I generally encourage parents to make as little physical contact with the child as possible during meltdowns (for many children, nothing turns a minor meltdown into a Chernobyl faster than physical contact).

Let's see what basket thinking is like when it goes well.

Drama in Real Life
Basket Weaving

I'd been working with Anthony for about a month, and his parents and I had come to the conclusion that his difficulty thinking through solutions to potential weather disasters was only the tip of the proverbial iceberg. Anthony had difficulty thinking clearly in all sorts of situations, driven by his anxiety, disorganization, and concrete style of thinking. We began using the baskets framework to help his parents clarify Anthony's true capabilities; identify their priorities; help Anthony hang in, so he could think more clearly when he was frustrated; remove UFOs (unnecessary frustrating objects) from his radar screen; and intervene primarily on the front end, rather than the back end.

"Tell me about these baskets my wife's been talking about," said the father, who had missed the previous session.

"The baskets help you and your wife decide which situations are important enough to be nonnegotiable—those are in what we call Basket A—which ones to remove from Anthony's radar screen altogether—those go in Basket C—and which ones on which it might be best to help Anthony learn to think things through when he's stuck—that's Basket B."

"So what should we put in Basket A?" asked the father.

"That's up to you and your wife," I said. "Safety is always in Basket A."

"What do you mean by safety?" the father asked.

"Hurting people—you, his sister—or destroying property," I said.

"He's more likely to destroy property than to hurt one of us," said the mother. "In fact, his sister handles him better than we do."

"Anything else in Basket A?" asked the father.

"Well, to help you decide, let's think about the Basket A litmus test. First, a behavior has to be important. Second, Anthony has to be capable of performing it consistently. And third, you guys have to be willing and able to enforce it. My sense is that, for Anthony, going to school is also in Basket A."

The parents agreed.

"What about Basket C?" asked the father.

"We've started putting a ton of things in Basket C already," said the mother. "We decided that his learning to tie his shoes could be taken off his radar screen for now, so we've been giving him a lot more help with that and some other things he gets frustrated about in the morning."

"Good," I said. "What else?"

"We've pretty much concluded that his doing a lot of his own writing is in Basket C for now, too," said the father. "We've been struggling to help him write for years, and he still clearly needs our help."

"That's fine," I said. "It sounds like you get the idea about Basket C. So let's think about Basket B."

"I had a few Basket B items this week," the mother said. "The other day he started freaking out when I told him to brush his teeth. I always thought it was because he just didn't want to do it. But I did the Basket B thing—I said to him, 'We want you to brush your teeth and you don't want to brush your teeth, so we need to come up with a compromise.'"

"What'd he say?" I asked.

"He said, 'Who says I don't want to brush my teeth?'" replied the mother. "I said, 'Because you always go nuts when I tell you it's time to brush your teeth.' He eventually told me the reason he didn't like to brush his teeth was because he had trouble putting the toothpaste on the toothbrush. You'd think I would have figured that out, with all his troubles with motor skills. So I asked him if he had any ideas about how we could compromise. It was actually his idea that I put the toothpaste on his toothbrush. Then he brushed with no problem."

"Fabulous," I said. "Isn't it great how he was able to come up with the compromise once you were able to help him stay coherent and tell you what was the matter? Not every child can do that this early in the process."

"So let me make sure I've got this straight," said the father. "He has trouble knowing what's bothering him and figuring out what to do. We help him get better at it, so he can eventually do it on his own. Some things he just has to do—like go to school—and others we don't care about anymore. Right?"

"Right," I said.

"So who's going to help him do this when he's twenty-five years old?" asked the father.

"Hopefully, you guys and his teachers will do such a great job of training him, he won't need this kind of help when he's twenty-five," I said.

"This is hard work, isn't it?" asked the mother.

"It's very hard work," I said. "But you're working hard already. Let's make sure you have something to show for all that hard work."

"How did he get this way?" asked the father.

"You know," I said, "Anthony loves predictability. And I think the reason he likes predictability so much is that he deals quite poorly with unpredictability, which requires a lot more flexibility and thinking. Our goal is to help Anthony get better at being flexible and thinking things through, so he can deal better with the unpredictable events that pop up in his life."

"How are the baskets different from what we're doing already?" asked the father. "When Anthony gets upset, I wait till he calms down and then I talk to him about how he could have handled the situation better. You don't think that's helping?"

"It might well be helping," I said. "But given that Anthony has a lot of trouble remembering a lot of the things he learns, I'm concerned that he doesn't seem to recall the things you've taught him the next time he becomes upset. I think the things we're trying to teach him are more likely to stick if your help is given when he needs it the most—in the early stages of his being stuck. Then he'll experience what it's like to figure out what he's anxious or upset about, how to think of solutions to problems, and how good you are at helping him. Ultimately—with lots of practice—I'm hopeful he'll start doing some of the calming, figuring out, and problem solving on his own."

Drama in Real Life
Mickey, Minnie . . . Meltdown?

Remember Casey from Chapter 3? He and his parents and sister took a trip to DisneyWorld, and their first day went wonderfully. They were a pretty tired*, hungry* crew as they left the Magic Kingdom on their way back to their hotel* (the asterisks designate well-known vapor-lock fuelers: fatigue, hunger, and transitions). Of course, this scenario had graver implications for Casey than for his sister. Just after they'd got outside the gates, Casey uttered the following, ominous request:

"I want cotton candy."

"You can't have cotton candy because we're not going back into the park to look for it," said the father instinctively.

Casey stopped dead in his tracks. "I want cotton candy!" he said loudly. Commence vapor lock.

The parents exchanged quick glances. They'd become pretty good at making quick "basket" decisions and went through their mental basket checklist quickly. This was not a safety issue, nor were any of their other few Basket A issues involved. Nonetheless, going back into the park for cotton candy would have been extremely inconvenient, and they did want Casey to eat a good dinner. Basket B or C?

"Look, Casey, we're all tired. . . ." The mother tried the empathy and logical persuasion route and hoped Casey still had a few coherent brain cells available. "Let's just go back to the hotel and get something to eat."

Casey was stuck. "I want cotton candy!" he said, moving closer to the edge of the cliff. Commence crossroads.

The parents exchanged glances again and decided to try Basket B. "I think we need to find a way to compromise on this, Casey," the father said calmly. "You want cotton candy, and we want to get back to the hotel to get something to eat. Can you think of a good compromise?"

"No!" Casey pouted, crossing his arms, still on the edge.

"Well, let's think about this for a second," said the father, crouching down next to, but not touching, his son. "We could look for someone selling cotton candy on the way back to the hotel . . . or we could buy you something to snack on besides cotton candy. Can you think of something else you'd like to eat on the way home besides cotton candy?"

"I want cotton candy," whined Casey, but his tone suggested slightly greater coherence.

"Well, I don't want to go back into the park, but I'm happy to look for someone selling cotton candy on the way home," said the father. Casey started walking toward the car again.

"Can I have cotton candy, too?" asked Casey's sister.

The mother bit her lip. "We'll get something we can all share," she said.

Once in the car, the family spent the next ten minutes with their faces glued to the windows, scanning the horizon for potential cotton candy vendors. Casey's coherence slowly returned. There was one small problem, of course: They hadn't yet stumbled across any stores that might be selling cotton candy. Nonetheless, with Casey's coherence restored, the father felt Casey might be able to handle the bad news without falling apart.

"I don't think we're going to find cotton candy, folks,"

he said, pulling the car into a McDonald's. "But let's see what kinds of snacks we can find at this McDonald's. Maybe some french fries."

"I want a Coke, too!" announced Casey.

"No, we're going to have to pass on the Coke right now," said the father, carefully reading his son's response. "You and your sister must have Coke coming out your ears by now. Let's hurry up and get something good."

Casey rushed into the McDonald's, ate his fries, forgot about the Coke, and wound up eating a decent dinner back at the hotel.

Had the parents decided that cotton candy was a Basket A issue, they would likely have endured a long meltdown. Had they decided it was a Basket C issue, they would have gone back into the park for cotton candy. By deciding the issue was in Basket B, they averted a meltdown on an important issue, and their son still ate some of what they wanted him to eat. It was their choice to make. They were in control of whether, and under what circumstances, they induced and endured a meltdown.

Drama in Real Life

Sitting on the Register Is Not in Basket A

Remember Helen, the girl who wanted macaroni and cheese instead of chili? One of the most difficult times of the day in Helen's family is when it's time for her to sit down and do her homework.

One night, Helen had somehow decided that she wanted to do her homework sitting atop the "register" in the kitchen. (I grew up in Florida, so Helen's parents had to explain to me that a "register" is a heating vent in the floor.) Helen's parents objected to her doing her homework sitting atop the register. This minor issue had the potential to disrupt Helen's completion of her homework by inducing a prolonged meltdown.

"Helen, you can't do your homework sitting on the register," said the father.

"I want to," whined Helen. Commence vapor lock.

"Helen, I want you to come over to the kitchen table and do your homework," commanded the father.

"I want to sit here!" Helen whined with a little more fervor. Commence crossroads.

Basket A, Basket B, or Basket C? Although the father felt strongly that Helen should do her homework at the kitchen table, he nonetheless quickly decided this was a Basket B issue (ideally he would have decided this at the first sign of vapor lock).

"Helen, we need to find some way to compromise on this issue. I don't want you doing your homework sitting on the register, and you seem to really want to do your homework sitting on the register. Can you come up with a compromise so that I'm happy and you're happy?"

"No, I want to sit here," Helen pouted.

"Come on, Helen, let's find a way for us both to be happy," the father encouraged.

"How 'bout I do my homework on the register tonight and at the kitchen table tomorrow night?" volunteered Helen, coherence kicking in.

"Well, that's an idea, but it wouldn't make me happy," said the father. "Can you think of some other way for us both to be happy?"

"No, that's it!" Helen responded.

"Come, on, I think we can do it together," the father said, gently pulling Helen away from the cliff. "How about you do your math homework on the register—as long as it doesn't take more than fifteen minutes—and your reading homework at the table."

"Uhm." Helen pondered this idea for a few moments. "OK."

"So if your math doesn't take longer than fifteen minutes, you can do your math on the register and your reading at the table. But at the fifteen-minute mark or when it comes time to do your reading, I don't want a big fuss."

"OK," said Helen, returning to her math assignment.

"You did a very nice job of compromising," said the father.

Homework done. Compromising skills reinforced. Relationship strengthened. And no meltdown.

In one of our next sessions, the father needed a little reassurance. "I'm afraid we're teaching her that she never has to listen to us, and I don't think that bodes well for the future."

"What, she never does what you tell her to now?" I asked.

"No, she actually does what we ask quite often," he replied. "I'm worried that she'll think that all she has to do is start to throw a fit to get what she wants."

"You've been doing the baskets for a few months now. Is she melting down less or more?" I asked.

"A lot less." The father smiled.

"Is she more or less compliant with Basket A issues?" I asked.

"More," said the father.

"Are you getting better compliance with things that aren't even in Basket A?"

"Yes, absolutely," the father replied.

"Are you yelling a lot less?"

"Yes."

"How are you and Helen getting along lately?" I asked.

"It's a lot better."

"Do you think that when she's in vapor lock, she's starting to be able to view you as a helper, rather than as the enemy?"

"Yes, I think so," the father replied. "You know, Helen was always a very affectionate kid. But we were battling so much that, up until a few weeks ago, when I'd get home from work she'd barely even acknowledge my presence. For the past few weeks, when I get home from work, she jumps up from whatever she's doing and gives me a big hug."

"I think we're doing OK," I said.

"But what about the real world?" the father asked.

"What about it?" I asked.

"The real world doesn't have baskets in it or people who always try to understand," he said.

"I know, but right now Helen seems to be doing pretty well in the real world. It's how she does at home that we're mostly concerned with. So we can let the real world take care of itself for now and focus on your relationship with your daughter. In fact, resolving disputes and working out disagreements is a much more important skill for her to learn for the real world than blind adherence to authority anyway."

Are there times when things don't go this well? Of course. Often they don't because some of the important ingredients are missing. For example, some parents forget the Basket A litmus test and continue to induce meltdowns on priorities that can't actually be enforced. One mother was determined that her eleven-year-old daughter should commit to wearing a life vest when she went boating with her friends. The mother had no way to prevent her daughter from going boating if she refused to wear a life vest, so the issue clearly wasn't in Basket A. Moreover, the mother had no way of enforcing her demand that her daughter wear a life vest. The issue failed to pass the Basket A litmus test, but this didn't prevent the mother from haranguing her daughter about the importance of wearing a life vest. It sounded something like this:

Mother: You'll be sure to wear a life vest when you're on the boat . . . because if something happened and you weren't wearing a life vest . . .

Daughter [*with slight agitation*]: I heard you, Mom.

Mother [*failing to read the slight agitation*]: Then you could find yourself hurt in the water and even though you're a good swimmer, well, if you got hurt it wouldn't matter how good of a swimmer you . . .

Daughter [*very agitated*]: I heard you, dammit!

Mother [*forgetting about the potential ramifications of her daughter's agitation*]: Don't get mad at me, I just want to make sure you're safe . . . it makes me very uncomfortable to know that you're out there . . .

Daughter [*incoherent, screaming*]: Get the fuck away from me! I hate your fucking guts! You are so fucking annoying! Leave me alone!

Mother: Leave you alone? You're lucky you have a mother who worries . . .

Kaboom. When I was told this story, it became clear that we had to take a few steps back. The mother needed more work on some of the components of the user-friendlier environment, including her reading skills. She also needed some refreshing on the Basket A litmus test.

Even when things do go well as you get better at using the baskets, there's a temptation to become complacent or even to increase sharply the number of priorities in Basket A. A simple word of advice: Don't. In the same way that former alcoholics try to take things one day at a time because they know that falling off the wagon is always one drink away, early on your child is still capable of melting down no matter how good you and he have become at compromising. It's like driving a car, in that you're always checking the gauges to make sure you have enough fuel, the engine isn't overheating, and you're not exceeding the speed limit. The same thing goes with your inflexible-explosive child: You've got to constantly check the gauges. Just because things are going well right now doesn't mean a problem isn't right around the bend.

Let's summarize the major points of this chapter.

- As part of creating a user-friendlier environment, you now have two major priorities: reducing the frequency of your child's meltdowns and helping your child to maintain coherence in the midst of vapor lock or to regain it during meltdowns.

- Thus, on the front end—before and during vapor lock and at the crossroads—you should ask yourself the following question: "*Is this behavior important or undesirable enough for me to induce and endure a meltdown?*"

- If the answer to this question is yes—and except for safety, it won't be often—the issue is in Basket A and is nonnegotiable.

Rather than putting your energy into thinking up better ways to reward and punish your child in the pursuit of improved compliance, you're putting energy into compromising and communicating for the purposes of rebuilding your relationship and setting the stage for improving the deficient skills underlying your child's inflexibility and low tolerance for frustration.

- If the answer to this question is no, but the issue is still a high priority, the issue is in Basket B, and you negotiate a mutually satisfactory compromise.

- If the answer is no and the issue is unimportant—and that's going to happen a lot more than it used to—the issue is in Basket C, and your lips are sealed.

- Basket A is important because it helps you maintain your status as an authority figure. Basket C is important because it helps you remove low-priority issues from the path of your child's cognitive wheelchair. But Basket B is the most important, since it provides a framework for your child to learn to stay coherent in the midst of frustration, take another person's perspective, engage in give-and-take, generate alternative solutions to a problem, think things through, and compromise. With your help.

Change can be slow, sometimes very slow, but you've put a lot of time in already. Rather than putting your energy into thinking up better ways to reward and punish your child in the pursuit of improved compliance, you're putting energy into compromising and communicating for the purposes of rebuilding your relationship and setting the stage for improving the deficient skills underlying your child's inflexibility and low tolerance for frustration.

8
Family Matters

An inflexible-explosive child can lay bare family issues that might never have risen to the surface had the parents been blessed with a less difficult child. But family issues can also complicate or impede the creation of a user-friendlier home environment. Maladaptive family communication patterns, for example, can make it a lot harder for family members to discuss important problems productively; in some instances, these communication patterns can actually fuel meltdowns. Sibling issues, never easy to deal with under the best circumstances, are even more troublesome when one of the siblings is inflexible and explosive. Sometimes parents have difficulties of their own—job stresses, financial problems, or marital issues—that may make it hard to devote extra energy to the creation of a user-friendlier household. And sometimes grandparents or other relatives don't make the task easier. We'd better take a closer look at these issues before assuming the smoke is fully cleared and the stage set for what comes next.

Siblings

Even in so-called normal families, adversarial interactions between siblings can be considered something of a rite of passage. But adding an inflexible-explosive child to the mix can make typical sibling rivalry look like a walk in the park. For example, though it's not uncommon for "normal" siblings to direct their greatest hostility and most savage acts toward each other, these acts can be more intense and traumatizing when they're inflicted—chronically—by an inflexible-explosive child in a state of incoherence. And though it's not unusual for "normal" siblings to complain about preferential treatment and disparities in parental attention and expectations, these issues can be magnified in families with an inflexible-explosive child because he requires such a disproportionate share of the parents' resources. Finally, though many siblings seem to get their thrills by antagonizing or teasing one another, an inflexible-explosive sibling may be considered less capable of responding to such antagonism in an adaptive way; such interactions may therefore have the primary effect of fueling countless meltdowns.

Fortunately, a user-friendlier environment can improve interactions among siblings. But siblings cannot merely be considered beneficiaries of such an environment; they are also essential to its creation. Therefore, depending on their ages, it is often useful to help brothers and sisters understand why their inflexible-explosive sibling acts the way he does, why his behavior is so difficult to change, how to interact with him in a way that reduces hostility and minimizes the likelihood of aggression or explosions, and what the parents are actively doing to try to improve things. Brothers and sisters tend to be more receptive if there's an improvement in the general tone of family interactions and if their inflexible-explosive sibling begins to blow up less often and becomes an active participant in making things better.

Nonetheless, this understanding doesn't always prevent siblings from complaining about an apparent double standard between them and their inflexible-explosive brother or sister. Armed with the knowledge that parental attention is never distributed with 100 percent parity and parental priorities are never exactly the same for each child, you should resist responding to this complaint by trying even harder to get your inflexible-explosive child to be like your other children. In all families—yours and everyone else's—fair does not mean equal. Even parents in "normal" families often find themselves providing one child with more help with homework, having higher academic expectations for one child, or being more nurturing toward another. In your family, you're doing things a little differently for the child who has deficits in the areas of flexibility and frustration tolerance. This can be a tough pill for brothers and sisters to swallow, so I often encourage parents to designate slots of private time with each sibling to compensate partially for the imbalanced distribution of parental attention. But when siblings complain about disparities in parental expectations, it's an excellent opportunity to do some empathizing and educating.

Sister: How come you don't get mad at Danny when he swears at you? It's not fair!

Mother: I know that it's very hard for you to listen to him swearing. I don't like it very much, either. But in our family we try to help one another and make sure everyone gets what he or she needs. I'm trying to help Danny stay calm when he gets frustrated and to help him think of different words he could use instead of swearing. That's what he needs help with.

Sister: But swearing is bad. You should get mad at him when he swears.

Mother: Well, I don't get mad at you when I'm helping you with your math, right? That's because I don't think getting mad at you would help very much. Remember how I used to get mad at Danny whenever he swore? It didn't work very well, did it? It just made things worse. So I'm doing something now that I think will eventually work better. I think it's already starting to work pretty well.

Sister: Will you help me think of different words if I swear?

Mother: Of course I would. Then again, you don't swear very much, which is really good. But if you were to start having trouble with swearing, I would try to help you stay calm and tell you different words, just like I'm doing with Danny. Luckily, I don't think you'll need much help with that.

Sister: Yes. Math is what I need help with.

Mother: Exactly.

How may basket thinking be applied to interactions between an inflexible-explosive child and his siblings? To begin with, safety is still in Basket A. Issues that don't matter much are still in Basket C. And compromising is still in Basket B. The only difference is that you're helping your children abide by this framework in their interactions with one another. Let's look at a few examples.

Brothers and sisters are often on the receiving end of the aggressive acts of their inflexible-explosive sibling. Thus, in some families, safety is of paramount concern. Because of Basket A, safety is now your family's top priority, and this can reduce the level of aggression dramatically. But in some instances—especially early in treatment—safety can be achieved only through skillful management of the home environment. For example, in some cases, I recommend that parents deprive siblings of the pleasure of one another's company unless they're under adult supervision. In more extreme cases, I'll suggest that it's too risky for siblings to be together at all. In such instances, siblings don't eat dinner together, don't watch television together, don't sleep in the same room together, don't sit together at the same table in a restaurant, don't sit near one another in the car, and so forth. Once parents become more confident that safety can be maintained, the never-together policy may be reconsidered.

Mother: But Danny's sister loves him so much. I can't imagine depriving them of each other's company.

Me: I can. Safety is more important than having Danny and his sister be together right now. It's our responsibility to make sure she's safe no matter how much she loves him.

Basket B is still where some essential negotiating, problem-solving, and cooperation skills are trained and practiced, except now these activities are also applied to sibling interactions. Thus, when disputes arise between an inflexible-explosive child and his brother or sister, Basket B is where parents help both parties stay coherent, articulate their concerns, frame the problem to be resolved, and work toward a mutually satisfactory solution. You're not acting as a referee, just as a facilitator of conflict resolution. The brothers and sisters end up feeling good because such disputes are less terrifying; they see that their views are being heard, are involved in the process of negotiating a solution that takes their needs into account, and see that you're able to handle things in an evenhanded manner. The inflexible-explosive child ends up feeling good because you've helped him stay coherent in the midst of frustration, prevented him from committing an aggressive act he'd be sorry for later, helped him negotiate a solution that takes his needs into account, and reinforced your role as a helper. Eventually, the goal is for them to work out their difficulties without your assistance, but that's probably a long way off. Once the smoke clears, it's often possible to begin having family meetings aimed at processing and resolving general issues, especially if your inflexible-explosive child is able to tolerate such meetings.

Unfortunately, I've seen instances in which the behavior of seemingly "angelic" siblings begins to deteriorate just as the behavior of their inflexible-explosive brother or sister begins to improve. This is often a sign that the emotional needs of the siblings require closer examination. In some cases, individual therapy may be necessary for brothers and sisters who have been traumatized by their inflexible-explosive sibling or who may be manifesting other problems that can be traced back to the old family atmosphere.

You may find that your family needs help working on these

issues; if so, a skilled family therapist can be of great assistance. You may also wish to read an excellent book entitled *Siblings without Rivalry*, by Adele Faber and Elaine Mazlish.

Communication Patterns

A family therapist can also help when it comes to making some fundamental changes in how you communicate with your child. Dealing effectively with an inflexible-explosive child is easier (not easy, easier) when patterns of communication between the parents and child are adaptive. When these patterns are maladaptive, dealing effectively with such a child is much harder. Many of the patterns depicted next have been described by a host of family therapy theorists and clinicians (for example, Drs. Arthur Robin and Sharon Foster in their book, *Negotiating Parent-Adolescent Conflict*). As you may expect, some of these patterns are more typical of older inflexible-explosive children. But the seeds may be sown early. Although not an exhaustive list, here's a sampling of some of the more common patterns:

Parents and children sometimes get into a vicious cycle of drawing erroneous conclusions about each other's motives or cognitions. I usually refer to this pattern as "speculation"; others have referred to it as "psychologizing" or "mind reading," and it can sound something like this:

Parent: The reason Oscar doesn't listen to us is that he thinks he's so much smarter than we are.

Now, it's not uncommon for people to make inaccurate inferences about one another. Indeed, responding effectively to these inaccuracies—in other words, setting people straight about yourself in a manner they can understand—is a real talent and requires some pretty complex, rapid processing. Of course, in an inflexible-explosive child, the demand for complex, rapid processing presents a problem—he's not very good at it. So while there are some children who can respond to speculation by making appropriate, corrective statements to set the record straight, an inflexible-explosive child may hear himself being talked about inaccurately and spin straight into vapor lock. This is an undesirable circumstance in and of itself, but it's especially undesirable because whether Oscar thinks he's smarter than his parents probably isn't worth spending much time discussing. In fact, this topic is a red herring that just distracts everyone from the main issue, which is that Oscar and his parents still haven't figured out how to get past his inflexibility and low tolerance for frustration so they can discuss and resolve important issues.

Of course, speculation can be a two-way street. From a child's mouth, it may sound something like this:

Oscar: The only reason you guys get mad at me so much is because you like pushing me around.

Such statements can have the exact same detouring effect, especially when adults follow the careening child in front of them straight through the flashing lights and detour barriers and right off the cliff:

Mother: Yes, that's exactly right, our main goal in life is to push you around. I can't believe you'd say that, after all we've been through with you.

Oscar: Well, what is your main goal, then?

Father: Our main goal is to help you be normal.

Oscar: So now I'm not normal. Thank you very much, dickhead.

Speculation is always a risky undertaking. So it's time for a user-friendlier rule: Each family member is allowed to comment only on his or her own thoughts and motivations. In other words, you should only speak for yourself, using a lot of "I" statements, such as "I worry about you getting to bed so late" or "I feel very hurt when you say that." If your child does need help articulating his needs or frustrations, your attempts to assist him should be framed tentatively ("Correct me if I'm wrong, but I think what you may be trying to say is . . ." or "Maybe what you're frustrated about is . . . ") and should involve an absolute minimum of psychologizing and value judgments. You're also going to need someone to keep conversations on track so they don't swing off the topic. Now, a therapist could be that someone for an hour or two a week. But ultimately, it's going to have to be a family member and, as you may expect, parents are the early front-runners for this position. In my experience, many of the explosions that occur in interactions with inflexible-explosive children have little to do with the issues that were the main topics of conversation in the first place. When issues are brought up in a way that doesn't elicit defensiveness, most of these children are willing or even eager to talk about important desired topics such as these:

• How they can handle frustration and think things through more adaptively and how you may be able to help.

• How you'd like to start trying to resolve disagreements in a mutually satisfactory manner through civil discussion.

• Things each family member is saying that make another family member defensive and how to speak to one another in a more productive way.

Of course, it's also critical to listen closely to what the child has to say on these topics; to recognize that it can take the child a long time to spit something out; and to remember that if a child isn't ready to talk about something at a given moment, you probably won't have much luck trying to force the issue. Come back to it later when you've got better odds.

Another maladaptive communication pattern has been called overgeneralization: the tendency to draw global conclusions in response to isolated events. Here's how it would sound from a parent:

Mother: Billy, maybe you can tell Dr. Greene why you're not doing your homework.

Billy: What are you talking about? I do my homework every night!

Mother: Your teachers told me you have a few missing assignments this semester.

Billy: So does everybody! What's the big fucking deal? I

miss a few assignments, and you're ready to call in the fucking cavalry!

Mother: Why do you always give me such a hard time? I just want what's best for you.

Billy: Stay out of my fucking business! That's what's best for me!

What a shame, because there may actually be ways in which Billy's mother could be helpful to him with his homework or, at least, get some of the reassurance she was looking for about his compliance with homework assignments. Not by starting the discussion with an overgeneralization, though. While other children are sometimes able to get past their parents' overgeneralizations to address the real issues, inflexible-explosive children often object almost immediately to statements that overgeneralize reality and may lack the skills to respond appropriately with information that corrects it. Phrasing things tentatively should help you overgeneralize less often and can therefore feel user-friendlier to your child ("*Oscar, I wonder if we can talk about this without screaming at each other*" or "*Billy, you'll let me know if there's anything about your homework I can help with?*").

In "perfectionism," parents fail to acknowledge the progress their child has made and demonstrate a tendency to cling to an old, unmodified vision of the child's capabilities. Perfectionism is often driven less by the child's lack of progress and more by the parents' own anxiety. Wherever it's coming from, perfectionism typically feels very user-*un*friendly to a child who may actually have been trying hard to stay on track or who may feel enormously frustrated by his parents' unrealistic expectations:

Father: Eric, your mother and I are pretty pleased about how much better you're doing in school, but you're still not working as hard as you ought to be.

Eric: Huh?

Mother: But that's not what we wanted to talk to you about. You're staying up too late doing your homework.

Eric: I get it done, don't I?

Father: Yes, apparently you do, but we want you to get it done earlier so you get more sleep.

Eric: I get enough sleep.

Father: We don't think you do. You're very grouchy in the morning, and you have trouble waking up. We want you to do your homework when you get home from school from now on.

Eric: I'm not doing my homework when I get home from school! I need a break when I get home from school! What fucking difference does it make?

Mother: It makes a difference to us. Now, your father and I have already talked this over, so there's no discussion on it. You need to get your homework done when you get home from school.

Eric: No fucking way.

Hmmm. Eric may or may not actually be interested in thinking about how to get his homework done earlier. Probably not. Either way, perfectionism is not a particularly effective way to engage him in a discussion on the topic. The ultimate antidote for perfectionism is perspective: Here's who my child was, here's who he is, and here's who he's likely to be—I should try to stop insisting that he become something he isn't.

Other maladaptive communication patterns include sarcasm, which is either totally lost on or extremely frustrating to inflexible-explosive children, who don't have the patience to figure out that the parent meant the exact opposite of what he or she actually said; put-downs (Parent: "*What's the matter with you?! Why can't you be more like your sister?*"); "ruination," sometimes called "catastrophizing," in which parents greatly overblow the effect of current behavior on a child's future well-being (Parent: "*We've resigned ourselves to the fact that Hector will probably end up in jail some day*"); interrupting (don't forget, the child is probably having trouble sorting through his thoughts in the first place—your interruptions don't help); lecturing ("*How many times do I have to tell you . . .* "); dwelling on the past ("*Listen, kid, your duck's been upside down in the water for a long time . . . you think I'm gonna get all excited just because you've put together a few good months?*"); and talking through a third person ("*I'm very upset about this, and your father is going to tell you why . . . isn't that right, dear?*"). All distinctly user-*un*friendly.

Over time, the goal is for you to be able to communicate with your inflexible-explosive child in a way that demonstrates to him that you can control yourself during discussions, stay on topic, recognize when discussions aren't going anywhere, get them back on track, and deal more adaptively with things that are frustrating to you both. This is very hard to do, and it's made even harder by the fact that you probably have powerful feelings of your own that influence the way you react to your child. But things can change; you may just need a little help to make them change. As a parent, you exert enormous influence over family communication patterns. Again, a family therapist can help you establish some basic rules of communicating that can improve interactions dramatically.

Over time, the goal is for you to be able to communicate with your inflexible-explosive child in a way that demonstrates to him that you can control yourself during discussions, stay on topic, recognize when discussions aren't going anywhere, get them back on track, and deal more adaptively with things that are frustrating to you both.

Parents

Needless to say, living with an inflexible-explosive child is a lot easier when adults communicate well enough to work as a team in establishing a user-friendlier environment. At the least, the adults need to reach a consensus on what belongs in Baskets A, B, and C. That means seeing eye-to-eye on the child's capabilities and agreeing on what's actually enforceable. If you're unable to reach such a consensus, your child will have to continue to handle two completely different sets of expectations, and we already know that boat won't float.

Now, while an inflexible-explosive child can put pressure on adults' relationships with each other, troubles between a couple can make life with such a child much more difficult to deal with. For partners who aren't even good at compromising with *each other,* reaching a consensus about the baskets and compromising with a child may be even more challenging. Partners who are drained by their own difficulties often have little left for a labor-intensive inflexible-explosive child. Sometimes one partner feels exhausted and resentful over being the "primary" parent because the other parent spends a lot of time at work. Power struggles that may occur between the adults often affect interactions with the child. And sometimes stepparent issues can enter the mix (Child: *"Stay out of this! You're not my real father!* or Stepfather: *"That kid was a problem before I arrived on the scene . . . this is between him and his mother"*). It's very difficult to make a user-friendlier environment work without addressing these issues. Sometimes marital and family therapy are necessary adjuncts.

Many parents feel very de-energized by their own personal difficulties. Some parents are bitter over having been dealt an inflexible-explosive hand by the great deck shuffler of children. For one mother, her son's explosions tapped into her own abusive childhood, and it was extremely difficult for her to get past her visceral reaction to her son's raised voice. Another simply felt so drained by being a full-time mother to her three other children that she simply had no energy left to help her inflexible-explosive son. A father had to get a handle on his own explosiveness before he could help his daughter with hers. Another father needed to be medicated for ADHD before he was able to stick to a plan agreed upon in treatment. Yet another father had to come to grips with his excessive drinking and its impact on the whole family before we could press ahead with creating a user-friendlier environment. It's difficult to work on helping your child if you're feeling the tug to put your own house in order first. Take care of yourself. Work hard at creating a support system for yourself. Seek professional help or other forms of support if you need it. These things don't change on their own.

Grandparents

At times it's necessary to bring grandparents into the therapeutic mix. In many families, grandparents or other relatives function as co-parents, taking care of the children while the parents are at work. If the grandparents aren't with the program, the inflexible-explosive child may be spending part of his day in a user-friendlier environment (when he's with his parents) and another part of the day in a user-*un*friendly environment (when he's with grandparents). This isn't an ideal scenario. In such circumstances, the grandparents are an integral part of the family unit and need to be brought into the conceptual loop (don't forget, it's important to ensure that all adults who interact with your child understand his unique difficulties). In other instances, the grandparents may not spend much time with the child but never miss an opportunity to remind the parents that what the child really needs is a good kick in the pants. These grandparents need to be brought into the loop as well, although sometimes they simply need to be enlightened about why their ideas probably aren't going to work with their inflexible-explosive grandchild.

Drama in Real Life:
Rules of Communicating

When Mitchell—the fifteen-year-old ninth grader who was diagnosed with Tourette's disorder and bipolar disorder—and his parents arrived for their second session, I was advised that it had been a difficult week.

"We can't talk to him anymore—about anything—without him going crazy," said his mother.

"THAT'S NOT SO, MOTHER!" Mitchell boomed. "I'm not going to sit here and listen to you exaggerate."

"Why don't you stand then?" the father cracked.

Mitchell paused, reflecting on his father's words. "If you were joking, then you're even less funny than I thought you were," he said. "If you weren't, then you're dumber than I thought you were."

"I'm not the one who flunked out of prep school," the father jabbed back.

"AND I'M NOT THE ONE WHO MADE ME GO TO THAT FUCKING SCHOOL!" Mitchell boomed.

"Look, I'm really not interested in getting into a pissing contest with you, Mitchell," said the father.

"What do you call what you just did?" the mother chimed in. "Anyway, I don't think Mitchell is ready to face flunking out of prep school yet."

"DON'T SPEAK FOR ME, MOTHER!" Mitchell boomed. "YOU DON'T KNOW WHAT I'M READY TO FACE!"

"Pardon me for interrupting," I said, "but is this the way conversations usually go in this family?"

"Why, do you think we're all lunatics?" asked Mitchell.

"Speak for yourself," said the father.

"Screw you," said Mitchell.

"Well, we're off to a wonderful start, aren't we?" said the mother.

"WE ARE NOT OFF TO A WONDERFUL START, MOTHER!" Mitchell boomed.

"I was being sarcastic," said the mother. "I thought a little humor might lighten things up a bit."

"I'm not amused," Mitchell grumbled.

"Fortunately, we're not here to amuse you," said the father.

"Sorry to interrupt you folks again," I said. "But I'm still wondering if this is a pretty typical conversation."

"Oh, Mitchell would have gotten insulted and stormed out of the room if we were at home," said the mother. "In fact, I'm surprised he's still sitting here now."

"YOU HAVE NO IDEA HOW I FEEL!" boomed Mitchell.

"We've been listening to you telling us how you feel since you were a baby," said the father. "We know more about how you feel than you know."

"ENOUGH!" boomed Mitchell.

"My sentiments exactly," I said. "I think I'll answer my own question. Forgive me for being so direct, but you guys have some not-so-wonderful ways of communicating with one another."

"How do you mean?" asked the mother.

"You're a very sarcastic group," I said. "Which would be fine, I guess, except that when you're sarcastic, I think it makes it very hard for Mitchell to figure out what you mean."

"But he's so smart and we're so dumb," said the father.

Mitchell paused, reflecting on his father's words. "Are you trying to be funny again?" he asked his father.

"You're so smart, figure it out," the father said.

"Uhm," I interrupted. "I'm sure you guys could do this all day, but I don't think it would get us anywhere."

Mitchell chuckled. "He still thinks we're going to accomplish something by coming here," he said into the air.

"I should add that sarcasm isn't the only bad habit," I continued. "The one-upmanship in this family is intense."

"Birds of a feather," the mother chirped.

"What does that mean?" Mitchell demanded.

"It means that the apple didn't fall far from the tree," said the mother.

"Be careful about whose tree you're talking about," said the father. "I don't want any credit for this."

"Oh, I'm afraid you're right in the thick of things," I reassured the father. "I wonder if we could establish a few rules of communicating. I must warn you, I'm not sure you'll have much to say to one another once I tell you these rules."

"Bravo," said Mitchell. "That's music to my ears."

"What kind of rules?" asked the mother.

"Well, it would be a lot more productive if we got rid of a lot of the sarcasm," I said. "It really muddies up the communication waters. And the one-upmanship has got to go."

The ensuing silence was broken by the father. "I don't think he can do it," he said, looking at Mitchell.

Before Mitchell could erupt, I interjected, "That's one-upmanship."

Mitchell's frown turned upside down. "Thank you," he said.

"This is going to be hard," said the father. "And no more sarcasm either?"

"Not if you guys want your son to start talking to you again," I said.

"Where's that team spirit, fellas?" the mother chimed in.

"That's sarcasm," I interjected.

"Ooo, this guy is tough," said the father, turning to his wife. "I don't like coming here anymore." He smiled.

"That's sarcasm, too," I said.

"My husband isn't accustomed to being corrected," said the mother.

"Oh, that reminds me of the last bad habit," I said.

"Oh, God, what did I say?" the mother said, covering her mouth.

"You guys talk for one another a lot," I said, "like you can read one another's minds."

"Well, we know one another very well," said the mother.

"That may be," I said, "but from what I've observed, your speculations about one another are often off-target, and they don't go over very well."

"What'd you call it?" asked the mother.

"Speculation," I said. "Thinking you know what's going on in someone's head. It just gets you guys more agitated with one another."

"No more speculation?" said the mother.

"No more speculation," I confirmed.

"What should we do if someone does one of those three things?" Mitchell asked.

"Just point it out to them without being judgmental," I said. "If someone is sarcastic, just say, 'That's sarcasm.'

If someone is one-upping, say 'That's one-upmanship.' And if someone is speculating, say . . . "

"'That's speculation,'" said Mitchell.

"My, we catch on fast," said the father.

"That's sarcasm," said Mitchell.

9
The Devil Is in the Details

I have good news and bad news. The good news is that creating a user-friendlier environment and implementing the baskets framework can have a dramatic impact on you, your child, and your family. The approach described in the last three chapters can help you move away from an adversarial relationship with your inflexible-explosive child while you maintain your role as an authority figure and help you remove many unnecessary frustrations from the path of your child's cognitive wheelchair, thereby greatly reducing the opportunities for vapor lock and meltdown and increasing the likelihood that your few remaining priorities will be achieved. It can also help you think more clearly when your child is in the midst of vapor lock and help you—not your child—determine whether the episode develops into a full-fledged meltdown. And it can help you demonstrate to your child that you understand how debilitated he becomes in situations requiring flexibility and frustration tolerance and hence enable you to help him maintain coherence in the midst of these situations so he can discuss and think through potential solutions.

Now for the bad news: You're probably not done yet. You see, we've still got some work to do on two of the most important components of a user-friendlier environment:

(1) identifying the specific situations that routinely and predictably cause your child significant frustration and (2) addressing the specific factors that contribute to his inflexibility and explosiveness. This chapter is devoted to an in-depth coverage of these two overlapping components.

Let's start with the first component. Taking a closer look at specific frustrating situations can help you anticipate, before vapor lock sets in, times when your child is going to need your help. So, assuming that your child isn't melting down every minute of every waking hour, we need to figure out when he *is* melting down. Do vapor lock and meltdowns occur primarily in interactions with certain *individuals*—mother, father, sibling, peer, teacher—in certain *settings*—home, school, soccer games, Cub Scout meetings—on certain types of *tasks*—reading, writing, playing Nintendo, getting ready for school, getting ready for bed—or during certain *times* of the day—early in the morning, upon arriving home from school, during homework, when the Ritalin wears off, during family meals, at bedtime, when he's bored? Are there certain topics, sounds, articles of clothing, or cognitive challenges that seem to induce meltdowns fairly regularly?

Let's assume you're able to identify certain situations that routinely and predictably cause your child significant frustration. Now you have to decide how you're going to approach each situation before your child starts to deteriorate, to see if you can set the stage for a productive outcome. You've got several options, depending on three important considerations:

- How important is it that your child successfully master the demands of the situation right now? In other words, is it a high priority?

- If it is a high priority, is your child realistically capable of mastering the demands of the situation right now? In other words, does he have the skills necessary to meet the demands successfully?

- If not, is it realistically possible to address the factors contributing to his difficulties right now to help him develop the skills that would make mastery of the demands of the situation possible? In other words, could he successfully meet the demands if we provided him with some extra help?

Your answers to these questions will help you decide how you want to approach each situation.

If you decide that your child's mastery of a given situation is not important, you're probably best off finding ways to help him avoid the situation, at least for the time being. There's nothing to be gained by inducing and enduring meltdowns over issues that aren't important.

If you decide that your child's mastery of a given situation is important right now, but that he's not yet capable of such mastery and it's not realistically possible to address the factors underlying his difficulties, your best option is to alter or make adjustments so as to make the situation easier for him to handle. There's no sense in asking your child to jump over a hurdle that you don't think he's realistically capable of clearing, no matter how important it is; it's better to lower the bar to a level just within his reach.

Finally, if you decide that it is important for your child to master the demands of a given situation right now and that, with some extra help, he is capable of developing the skills to deal with the situation more effectively, your best bet is to provide additional training and support to enhance existing skills or teach those he lacks.

Let's take a closer look at each possibility.

Not Important (Right Now)

An example of the first possibility is the mother who found that her moody, impulsive, hyperactive six-year-old son, Eduardo, routinely melted down when she brought him to the supermarket. Eduardo melted down in other situations as well, of course, but none as predictably as the supermarket. Maybe it was the overstimulation, maybe it was the fact that he had very inflexible ideas about the foods he wanted his mother to buy (most of which were not at the top of his mother's list). Whatever the reason, no matter what the mother tried—preparing him for trips, rewarding him for good behavior and punishing him for inappropriate behavior, making shorter trips, having Grandma accompany them, trying to steer him around the aisles where meltdowns seemed to occur most often, agreeing that he could select one or two of the foods on his list—he still routinely melted down when she brought him to the supermarket. The mother finally came to the conclusion that mastery of the demands of the supermarket—staying next to the shopping cart, not demanding the purchase of every high-sugar cereal on the shelves, being patient in the checkout line—wasn't especially important at that point in her son's development. She decided he'd be much better off if she went to the supermarket without Eduardo.

Mother: But he can't avoid supermarkets forever, right?

Me: Right. Luckily, going to the supermarket is not critical to Eduardo's existence right now.

Mother: When should I try taking him into supermarkets again?

Me: When it becomes more important, and you think he can do it.

Mother: It's not always easy for my mother to watch him for me.

Me: I know. But it's even harder—and a lot more detrimental to your relationship with your son—to have him melting down every time you take him to the supermarket.

Important but Neither Capable nor Possible to Address (Right Now)

Homework is often a good example of the second possibility—important but not capable and not possible to address. Many parents, teachers, and school administrators believe that homework is an essential component of a child's education. Which is fine, except that many inflexible-explosive children find homework to be remarkably frustrating because they don't have any brain energy left after a long day at school, their medication has worn off, they have learning problems that make completing homework an agonizing task, or because homework—especially long-term assignments—requires a lot of organization and planning. Thus, it's no accident that these children often exhibit some of their most extreme inflexibility and explosiveness while they are trying to do homework.

Do these difficulties render some children incapable of completing the same homework assignments as their classmates? Yes. Is it always possible to address these difficulties effectively? No. Does having a child melt down routinely over homework help him feel more successful about doing homework? No. Are these difficulties a good reason to alter or adjust homework assignments? Yes. I've yet to be convinced that the best way to instill a good work ethic in a child—or to help his parents become actively and productively involved in his education—is by inducing and enduring five hours of meltdowns every school night. The best way to instill a good work ethic is to assign homework that is both sufficiently challenging and doable in terms of quantity and content. Achieving this goal, of course, takes a little extra effort by the adults who are overseeing the assigning and completing of homework.

How can homework be altered or adjusted in a manner that brings it into the doable range for an inflexible-explosive child? Probably in the same ways that are useful for children with other types of disabilities. I've found that once teachers are made aware of the horrific battles that are taking place between parents and children over homework they've assigned, most are willing to make all kinds of adaptations to homework assignments for their inflexible-explosive students. Good teachers know that although it would be nicer and more efficient to have all the students in a classroom have the exact same learning styles and capabilities, it never works out that way. So it's always necessary to adapt lessons and assignments to individual learners. Inflexibility-explosiveness is as good a reason to adapt schoolwork as is any other type of learning disability. Thus, when warranted, teachers usually agree to eliminate unnecessary repetition from homework assignments (such as permitting a student to do five multiplication problems instead of twenty); to place limits on the amount of time a child spends on homework each night; and to establish priorities among assignments, so the more important ones are completed first, and the child isn't penalized for not finishing the less important ones.

I've also found that most teachers are willing to allow students to produce work in a manner that keeps them from becoming tripped up by their learning difficulties. For example, if a teacher wanted to have students practice writing skills and an inflexible-explosive student became extremely frustrated by the physical demands of writing, the teacher might permit the student to complete his work on a word processor or by dictating his thoughts to a parent, sibling, or fellow classmate. Although it would be important to make sure that the student continued to work on writing skills, it would be even more important to make sure that our expectations took into account both the student's writing capabilities *and* level of frustration.

Finally, it's crucial for parents and teachers to communicate clearly and continuously about homework, so the teacher is well aware of the assignments that routinely frustrate the child. If a teacher isn't aware of a child's frustration on specific tasks, she or he can't be expected to make appropriate adjustments.

These steps often make homework more doable for many

inflexible-explosive children. But, in some cases, the battles over homework persist even though homework has been made user-friendlier for the child. Sometimes these continued battles are a signal that the parents are not, and may never be, the ideal homework monitors. In such instances, I often recommend that the child complete homework either at school, under the supervision of an adult whose feedback and assistance is less objectionable to the child (many schools have "homework clubs" that can serve this purpose), or at home, with the assistance of a high school–aged "homework helper" or hired tutor. Such assistance often improves communication between the parents and child and permits the parents to focus on the more important aspects of parenting.

Parent: What does it say about me that I can't help my own kid with his homework?

Me: Oh, I think it's entirely possible to be an outstanding parent and not be the one who helps your child with his homework.

Parent: But his teacher seems to feel that homework is my responsibility.

Me: Does the teacher know what goes on in your home during homework?

Parent: Not exactly—it's a little embarrassing.

Me: I know—but I think it's time to bring the teacher into the loop.

Parent: But what are we going to do, adjust his homework forever?

Me: I don't know. It's possible that the adjustments will set the stage for him to feel a greater sense of mastery around homework. Then fewer adjustments may be necessary. It's also possible that he'll always need some adjustments to his homework no matter how successful he feels.

Parent: But his future employers aren't going to make these kinds of adjustments for him.

Me: My bet is that he'll get a job doing something he enjoys more and is better at than the homework he struggles over. Anyway, thinking about his future employers while he's in the midst of being frustrated about his homework probably doesn't help you respond optimally. I think if we continuously focus on making things go well in the present, the future will take care of itself.

Important and Capable, with a Little Extra Help

Finally, let's say that your child's mastery of the demands of a particular situation is important and that he realistically seems capable if he is given additional training to enhance his existing skills, acquire the skills he lacks, or is helped to develop compensatory strategies. Rule number one: Look before you leap. Just because your child occasionally demonstrates the capacity to deal with the demands of a given situation doesn't mean he can do so consistently, so be careful about jumping to conclusions about his capabilities. Also be careful about deciding that *your* child is truly developmentally ready to master the skills demanded by a particular situation just because most of the *other* children his age seem ready. Your child may be on a different time line, and though it can make life harder for him and you, it's still not a criminal offense in most states. Finally, if your relationship with your child still involves a lot of arguing and hostility, you're probably not in an ideal position to assume the role of skill trainer just yet. Indeed, one of the most effective ways to increase arguing and hostility is to routinely overestimate your child's capacity to master a given task or continuously impart conventional wisdom like, "You could do it if you really put your mind to it," "You're just going to have to bite the bullet," or "I guess you're just going to have to hit rock bottom before you're ready to try."

Rule number two: Be skeptical. In many domains, the theoretical appeal of skills-training hasn't panned out too well in actual practice. The equivocal research on skills-training is probably due to a few factors. First, it's very hard for a child to practice and maintain new skills if everyone around him (parents, teachers, coaches, peers) is still behaving the same old way. If you really want a child's new skills to endure beyond the confines of his therapist's office, important people in the outside world are going to have to participate in helping the child learn and practice the skills. Second, some children never become great readers or develop great written-expression skills, no matter how badly adults want them to. Some child psychologists never become good enough at basketball to replace Michael Jordan. Our goal, of course, is realistic progress.

But let's assume that your child could, indeed, benefit from some skill enhancement to help him deal more adaptively with situations that routinely frustrate him. What skills might be worth training? That depends on the specific factors underlying his difficulties. Remember, it's going to be difficult to address your child's needs if those needs aren't well understood. You don't want to try to train your child in skills of which he is presently incapable or to waste time training him in skills he's already mastered. That's why the comprehensive assessment described in Chapter 3 is so important.

The next sections describe some of the skills I find lacking in many of the children with whom I work, along with some of the strategies I've found useful for addressing the children's difficulties.

Social Skills

One area in which many inflexible-explosive children need some additional training is social skills. As fate would have it, social skills tend to be pretty tough to train. Most of the children I work with are good enough at the simpler, more concrete aspects of social functioning, like awareness of body space, starting a conversation, entering a group, and maintaining eye contact. Many run into trouble on the complex aspects of social interactions, such as empathy, perspective taking, understanding the impact of their behavior on other people, and being aware of the way in which their behavior causes them to be perceived by others.

A nine-year-old boy with whom I was working—he'd been diagnosed with ADHD and had some mild nonverbal difficulties—was doing a remarkable job of alienating his classmates at school and his parents and younger brothers at home. Like many children with ADHD, Dennis tended to interrupt conversations and intrude into others' space. He would also say things impulsively that were construed as mean. Dennis received plenty of feedback and punishment for this problem, with little sign of improvement; indeed, the primary effect of these responses was the inducement of meltdowns. Medication was helping with a lot of his other ADHD-related behaviors, such as paying attention and concentrating, completing assignments, and being less motorically restless, but these social difficulties remained. He wasn't happy about this state of affairs but didn't seem able to do anything to help himself. Just as some people have a poor sense of direction and need a road map to help them find their way, Dennis seemed to have a poor sense of his effect on other people and needed a social road map.

Me: Dennis, I think we need to help you stop saying things to people that they think are mean.

Dennis: I know.

Me: Because I don't have the impression that you're a mean kid.

Dennis: Yep.

Me: Do you know the things you're saying may be hurtful before you say them?

Dennis: They sorta just come out. I don't really think about whether they're mean. I remember they're mean when kids get mad at me or I get into trouble.

Me: Can you think of any reason you'd want to say things that would hurt people's feelings? Like, are you trying to get the attention of the other kids?

Dennis: No.

Me: Are you saying these hurtful things because you like getting into trouble?

Dennis: No.

Me: Maybe you like having the other kids be mad at you?

Dennis: No way!

Me: Aren't you in a social skills group at school?

Dennis: Yep.

Me: Is the group helping you learn more about getting along with other kids?

Dennis: Yep.

Me: So how come you're still having trouble getting along with other kids?

Dennis: I don't know.

Me: Is this something you'd like some extra help with?

Dennis: Yep.

The consensus among Dennis, his parents, and me was that it was important for Dennis to get along better with his peers and family members (right now) and that with a little extra help, he was capable of doing it. We also agreed that Dennis was kindhearted and that his mean comments had less to do with maliciousness and more to do with poor impulse control and a possible lack of appreciation for how certain comments may cause others to feel. This being the case, we had little faith that the parents' usual way of responding ("That was a terrible thing to say! Go to your room and don't come out until you're ready to start thinking about how you're making other people feel!") was going to be helpful anytime soon.

First, we had to put some time into clearing the smoke. Over the years, interactions between Dennis and his parents had become decidedly adversarial. Dennis wasn't going to be receptive to the parents' help until the general tone of these interactions improved. So the parents spent five or six weeks working on creating a user-friendlier environment (in some families, this endeavor takes a lot longer). Once this environment was achieved, we began focusing on helping Dennis develop a better sense of why certain things he said might be hurtful. We limited our efforts to the home environment first and agreed, with Dennis's input, to have his parents begin giving him feedback—in a calm, user-friendly manner—about how his comments made them feel. Rather than get angry and punish him when he said hurtful things, his mother and father would instead call his attention to the fact that his words were hurtful and why what he said caused them to feel that way. It sounded something like this:

> Father: Uhm, Dennis, when you tell me my tie is ugly, it makes me feel bad.
>
> Dennis: Oh, yeah.
>
> Father: You don't have to like my tie, it's just that it makes people feel bad when you tell them something they're wearing is ugly.
>
> Dennis: I know.
>
> Father: I had a feeling you probably knew.

The parents gave Dennis similar feedback when they heard him speaking to his siblings in a hurtful manner and gave his siblings a few pointers on doing the same. After three or four weeks, we were convinced that Dennis was starting to become pretty clear about the kind of comments that might be hurtful to others. Indeed, with this kind of feedback being given on a fairly continuous basis, considering the impact of his words before he uttered them began moving to the front burner in Dennis's brain. The parents noticed that Dennis would occasionally stop in midsentence when he was making an unkind comment and sometimes would apologize if he couldn't stop himself in time.

There were several more items on our agenda: expanding

our efforts to the school and helping Dennis begin to understand how his words caused him to be perceived by other people. Dennis agreed that his teacher should be brought into the loop, and the parents then spoke to her about what we'd been up to. She was happy to join in on the feedback and reminding and used a variety of subtle ways to remind Dennis of his goal. The parents continued providing Dennis with feedback about how his words made them feel (so the original goal stayed on the front burner in Dennis's brain), but they also began providing Dennis with feedback about how his words and actions might cause others to perceive him.

"Dennis, when you tell other children their answers are dumb it probably makes them think you don't care how they feel. I think you do care how they feel, so I don't think you'd want to say things that give other kids the wrong idea."

Does Dennis still slip every so often? Absolutely. Does he still need a little feedback and reminding? Yep. Are things a lot better than they were? Without question. Would this progress have occurred if his parents were still engaged in an adversarial relationship with their son? I doubt it. Dennis was already motivated to do well; he just needed a little help.

Inflexible-explosive children often need help with another aspect of social functioning: their interpretations of things that happen to them and the way others feel about them. Many of the concrete children with whom I work have ways of interpreting things that happen to them—sometimes referred to as "attributions"—that are rote and often inaccurate. The same inflexibility that typifies their interactions with the world also colors their interpretations of those interactions and can have adverse effects on their mood and self-esteem. These cognitive biases tend to be both obvious and tough to change.

Me: Cindy, how do you like school?

Cindy: I hate school.

Me: You hate school? What is it about school that you hate?

Cindy: I just don't like it.

Me: That's a shame because you have to spend a lot of time there. But what is it that you don't like?

Cindy: The other kids think I'm stupid.

Me: They do? How so?

Cindy: They just do.

Me: Tell me what you mean by "stupid."

Cindy: You know . . . dumb . . . stupid.

Me: That must not feel very good to you. What makes you think the other kids think you're stupid? Do they say you're stupid?

Cindy: No . . . not exactly. I just know they think that.

Me: Is there anyone else who makes you feel that you're stupid?

Cindy: No.

Me: Well, there must be some reason you think the other kids think you're stupid. What made you decide that?

Cindy: I sometimes say the wrong answer in class.

Me: That can be very embarrassing. Can you remember a time when that happened?

Cindy: Well, like last week we were doing math problems on the chalkboard, and my answer was wrong.

Me: Did the other kids laugh?

Cindy: No . . . not really.

Me: Did someone tease you about putting the wrong answer?

Cindy: No.

Me: Did any of the other kids put the wrong answer?

Cindy: Yes. . . . Lots.

Me: Were they stupid when they put the wrong answer?

Cindy: No.

Me: What's different when you put the wrong answer?

Cindy: I don't know. I just know they think I'm stupid.

Me: Hmmm. When you put the wrong answer, you're stupid, but when they put the wrong answer, they're not. I don't get it.

Cindy: Me either.

Me: It's a little confusing because you get really good grades. How can you be stupid if you get really good grades?

Cindy: I don't know.

Me: Maybe you're not stupid.

Cindy: No, I'm stupid. Like sometimes I don't understand what I read right away and I have to go back and read it again.

Me: That happens to me sometimes, too. It's interesting, I don't think I'm stupid because I have to read things twice.

Cindy: That's you.

Me: Does your mom know the other kids think you're stupid?

Cindy: Yes.

Me: What does your mom say about that?

Cindy: She tells me she thinks I'm very smart.

Me: Does that help you think you're not stupid?

Cindy: No.

Me: Is there anything else about school you don't like besides that the other kids think you're stupid?

Cindy: Well, nobody likes me.

Clearly, such inflexible interpretations—"I hate school," "The other kids think I'm stupid," "Nobody likes me," and so forth—often defy logic. And in many instances these interpretations contribute to a child's cumulative level of frustration or fuel his frustration at a given moment. Sometimes the hard part is to distinguish such statements from mental debris. In general, when these statements are made only in the context of vapor lock and meltdown, they're usually just a sign that the child is having trouble thinking clearly *at that moment*. If they occur outside the context of vapor lock and meltdown, the statements may reflect something less fleeting.

Entire books have been written on how to "restructure" the inaccurate, maladaptive thoughts of children and adults. The idea is to help the individual recognize the inaccuracy of his existing belief systems (sometimes called "schemas") and replace the inaccurate thoughts that make up these belief systems with more accurate, adaptive ways of thinking. This restructuring usually involves "disconfirming" the individual's old thoughts by presenting—in a user-friendly, low key, systematic manner—evidence that is contrary to these rigid beliefs. With a child who is stuck on the belief that she's stupid, we might have a teacher or parent whisper the following comment in response to a good grade on an assignment: *"I know you sometimes think you're stupid, but I don't think someone who's stupid could have done that well on that task."* In a child who has bona fide weaknesses in one area and strengths in another, a teacher's feedback might be as follows: *"I know you're struggling with reading—and that makes you say you're stupid sometimes—but I've never seen anybody who was so good at math. It would help me a lot if you'd help some of the other kids in our class who aren't so good at math."* With such feedback being presented continuously, we can, over the long haul, sometimes make a dent in a child's concrete belief system. Parents and teachers do so with *all* children. It just takes on a little more urgency and requires more time, patience, and hard work with inflexible-explosive children.

Recognizing, Expressing, and Thinking Through Frustration

As you already know, the angry, accusatory, disrespectful, personalized way in which many inflexible-explosive children express their frustration is sometimes due to their difficulty recognizing, expressing, and thinking through frustration. Clearly, these problems—which can often be traced to deficits in language processing, executive functions, and nonverbal skills—are important, and in many instances these children are actually capable of responding more adaptively with a little help, sometimes from a speech and language therapist.

Although this sequence of events—first *recognizing* that one is frustrated, then *expressing* that frustration, and then *thinking through* (and perhaps discussing) effective ways of handling the frustration—comes naturally to many people, it can be incredibly difficult for children who are executively, nonverbally, or linguistically impaired because it requires organization, planning, working memory, separation of affect, problem solving, flexibility, and expressing oneself and thinking via language. Breaking down the process of dealing with frustration into its components, providing the child with a simple framework for applying these parts, and helping the child with the parts he's not so great at can make things more manageable. Let's think about these components one at a time.

As strange as it may seem, some children actually have trouble recognizing that they're frustrated or that they're experiencing other feelings—such as fatigue or hunger—that may lead to frustration. It's hard for a child to figure out what to do about being frustrated if he doesn't realize that's what he is. A user-friendlier environment should give you a good jumpstart on helping such a child because you're already reading the early warning signs of frustration and providing the child with feedback about these signs ("Casey, you're starting to look a little frustrated, is there something I can help you with?"). Nonetheless, we eventually want the child to recognize that he's frustrated on his own.

Some children benefit from being helped to recognize the physical reactions that can accompany frustration (such as fatigue, hunger, a hot face, burning ears, dizziness, and head spinning). But I often provide such children with a rudimentary vocabulary for categorizing their feelings and try to help them recognize when they are experiencing these feelings. Often, this vocabulary consists of three feelings: happy, sad, and—of course—frustrated.

We helped Helen, a child you read about in Chapter 4, learn and practice her rudimentary vocabulary by having her parents discuss the past day's events with her at bedtime in a relaxed, user-friendly, nurturing manner. They would ask her what happened during the day that made her "happy," whether anything made her "sad," and whether anything made her "frustrated." If Helen couldn't remember specific events that fit one of these three categories, her parents would suggest some possibilities. If Helen had difficulty labeling the emotions that were associated with a particular event, her parents would help her. The parents made sure Helen's teacher was also aware of her new vocabulary so Helen didn't become confused by different terminologies. Over the course of several months, Helen slowly began to express her emotions across an increased number and range of situations. Her vocabulary of emotions became broader and more sophisticated and included such terms as *confused, disappointed, excited, bored,* and *annoyed.*

As you may imagine, "I'm frustrated" is a lot more palatable to most parents than "Leave me alone!" or "Go to hell!" and makes it a lot easier to be a helper. For the linguistically compromised child, this vocabulary provides a simple language for expressing emotions. For the executively compromised child, it requires minimal processing and imposes an organizational framework. For the child with a nonverbal learning disability, the use of language to express frustration is just what the doctor ordered.

Another inflexible-explosive child—a boy named Trent, who was diagnosed with bipolar disorder—was involved with a social service agency, and the social worker assigned to the family, Julie Golden, was superbly attuned to Trent's difficulties. One afternoon, she and Trent collaborated on a rating scale to help him develop an awareness of his different levels of frustration and the behaviors associated with each level. Here's the scale they came up with:

LEVEL	COLOR/EMOTION	BEHAVIORS
10	Bright red	Totally out of control (throwing things, breaking things, body out of control)
9	Medium red	Very out of control (running away from Mom and Dad)
8	Light red	Pretty out of control ("I don't want to live here!")
7	Bright Orange	Mouth out of control ("I don't care")
6	Medium orange	Jumping up and down (yelling "No!")
5	Light orange	Fresh talk (saying "No" to Mom and Dad)
4	Bright yellow	Getting "antsy," bored (bugging sister, being aggravating)
3	Medium yellow	Tired, cranky (complaining)
2	Light yellow	Getting tired (losing patience)
1	Green	Controlled, calm, good mood

Trent and his family found this scale useful for recognizing and communicating that he was frustrated.

Many children get the basic vocabulary down but still fall apart because they can't figure out or tell people what's the matter. So while they may recognize that they're frustrated—and may even be able to say "I'm frustrated"—they become cognitively paralyzed the instant they try to think about or describe what it is they're frustrated about. Such children may need additional help figuring out what they're frustrated about and finding the words to articulate their frustration with expressions like, "That's bothering me," "I don't know how to do that," "I don't know what to do," "I'm scared," "I don't feel right," "I need a break," and "I need some help."

Of course, you'll need to remind your child of his new vocabulary and respond in a way that encourages its continued use. Some children take slowly to their new vocabulary and continue to express frustration in ways (*"No! I can't do that right now!"*) that still require enlightened interpretation and a gentle reminder (*"Boy, you sure sound frustrated!"*). In some instances, children begin to verbalize their frustration more adaptively, but adults fail to appreciate the significance of this skill (*"You're frustrated? Just do what you're told, and you won't be frustrated!"*).

Father: We had a few major blowouts this week.

Me: Over what?

Father: I don't like the way Andy speaks to me when I tell him to do things.

Me: Can you give me an example?

Father: Well, this is a pretty common scenario. He's watching TV the other morning, and I tell him he needs to put his shoes on. He tells me to leave him alone.

Me: What did you do?

Father: I screamed at him.

Mother: See, I think the problem was that he usually puts his shoes on after he eats breakfast.

Father: That doesn't give him the right to speak to me that way.

Me [*to Andy*]: Andy, do you remember saying "Leave me alone" to your father?

Andy: Yes.

Me: What would cause you to tell your father to leave you alone after he'd told you to do something?

Andy: I was mesmerized.

Me: Mesmerized?

Andy: Yes, when I watch TV, I get mesmerized.

Me: So it was hard for you to drag yourself away from the TV to put your shoes on?

Andy: Kinda.

Me: Andy, can you think of other things you could say to your parents when they ask you to do something that they don't like?

Andy: Yes. Sometimes I say, "I don't want to."

Father: Sometimes you just yell, "No!"

Mother: Sometimes you say, "I don't have to listen to you" or "You can't tell me what to do."

Me: What's our explanation for why Andy says these things?

Father: I think he's just very disrespectful sometimes.

Me: May I offer a different explanation?

Father: Sure.

Me: Based on what we know about Andy—his poor impulse control, distractibility, and difficulty with expressive language—I'm wondering if maybe he isn't that great at shifting gears and at articulating the frustration he feels when he's asked to make such shifts.

Father: You lost me.

Me: Here's what I mean. Andy, what could you have said to your father when he asked you to stop watching TV and put your shoes on?

Andy: Good morning?

Me: Hmmm. You could have said that. But what were you trying to tell your father?

Andy: That I wanted to watch TV longer.

Me: How could you have said that?

Andy: I don't know.

Father: I think I see what you mean. So what should I have said?

Me: Well, it depends on what you wanted to teach Andy. I must confess, I can't think of any lesson that would be best taught by your screaming at him.

Father: I just can't tolerate disrespect. My immediate reaction is to start screaming.

Me: What I'm saying is that I think Andy has the utmost respect for you already, particularly in his coherent moments. Do you agree?

Father: In his coherent moments, yes.

Me: So there must be some other reason Andy speaks to you in a disrespectful manner when you ask him to do things when he's watching TV.

Mother: You're saying he says these things because he doesn't know how to say he's frustrated any better?

Me: Precisely.

Father: So what should we say to him when he says these things?

Me: Well, there are different ways to do this, but ultimately your goal is to give him better words.

Father: Give me an example.

Me: Well, let's take the most recent episode. You told Andy to put his shoes on while he was watching TV. He responded with "Leave me alone." He probably wanted to be left alone, so his words, while disrespectful, were probably right on target. But instead of thinking, "I will not be spoken to that way!" and getting all heated up, I think you'd be a lot better off thinking, "I bet I can help him say that differently" and giving him different words like, "You'd like to watch TV a few more minutes?"

Andy: I could have told him that Mom always lets me put my shoes on after breakfast.

Me: You could have . . . why didn't you?

Andy: I was mesmerized. And he started yelling too fast.

Me: Andy, is there something your mom and dad could do to help you get unmesmerized when they want you to do something while you're watching TV?

Mother: I'll tell you what I do . . . I go in and rub his shoulders a little . . . that demesmerizes him.

Me: Then he's better able to respond to you in a respectful way?

Mother: Absolutely.

Father: This is going to be very hard.

Me: I know. But think about how hard you're working already. Let's at least make sure you have something to show for all that hard work.

With lots of help from his parents, the manner in which Andy expressed frustration and asked for help became more sophisticated. For example, when he was in "information overload," he began saying things like, "Slow down, I can't keep track." When he became confused, he'd say, "I don't know what to do." When he felt himself becoming frustrated and needed some help, he'd say, "I feel really bad right now." And when he needed some extra time to organize an appropriate response (and avert an inappropriate one), he'd say, "I think I need a minute here."

Swearing in the midst of frustration is often a clear sign that a child doesn't currently have the thinking or linguistic skills to express frustration adaptively. With this interpretation and explanation of swearing in mind, our main goal is the same: to help the child alter his method of delivery and use different words. Responses like, *"I refuse to be spoken to like that!"* or *"Go to your room and come back when you're ready to talk to me the right way!"* may be close to the mark if your child lacks the knowledge that you don't want to be spoken to that way or isn't motivated to speak to you the right way. But if the true issue is that your child lacks the thinking or linguistic skills—or coherence—to articulate frustration, these comments are way off the mark.

Mother: We had a bad incident this week, and I'm not sure I handled it well.

Me: Tell me.

Mother: Well, I was making pancakes for breakfast. Derrick came into the kitchen and said he didn't want pancakes. I told him that's what was on the menu. . . .

Me: Sorry to interrupt, but was eating pancakes in Basket A?

Mother [*smiling*]: No.

Me: Just curious. Go on.

Mother: So then he called me a "fucking bitch" and ran

out of the kitchen. I ran after him and told him he was grounded for a week for calling me a name. He told me to get away from him. I insisted on an immediate apology. He went ballistic for the next half hour.

Me: It sounds extremely unpleasant. You mentioned that you wished you'd handled things differently?

Mother: First off, since eating pancakes wasn't in Basket A, I could have helped him find something else to eat.

Me: True. . . . Baskets B and C can spare him and you a lot of meltdowns. Anything else?

Mother: I guess I shouldn't get so upset when he swears at me.

Me: It's very hard not to get upset when your son calls you a fucking bitch. But you've been punishing him for calling you names for a very long time, and he still calls you a name every time he gets frustrated. So I don't think he needs any more lessons on the importance of not swearing or any more motivation not to swear. At the moment, I don't think Derrick is capable of expressing his frustration without swearing. In fact, since we've forbidden hitting—because it's unsafe—my sense is that, at this point, swearing is the only way Derrick has to express his frustration. Now, we've started trying to give him different words to use, but I don't think the new words come to him automatically yet. So I suspect we've got some more swearing to endure. The truth is, each time he swears, he gives us an opportunity to help him find different words.

Mother: So what should I have said?

Me: If eating pancakes was in Basket C, you could have said, "I think what you're saying is that you'd like something else to eat besides pancakes. Let's see if there's something else you'd like." If the issue was in Basket B, you could have said, "I think what you're saying is that you're frustrated about the pancakes—let's talk about that so we can figure out a good compromise."

Mother: So I blew it, right?

Me: No, you didn't blow it . . . this is really hard. Your understandable knee-jerk instinct is to punish your kid when he swears at you, and I'm asking you to do something different because punishing him hasn't worked. In other words, I'm asking you to hold on to your knees. It's hard to do.

Mother: Yes it is.

Me: I think it's worth pointing out that Derrick did something very adaptive in the incident you described.

Mother: He did? What?

Me: Instead of going toe-to-toe with you in the kitchen, he detached himself from the situation. In other words, he left. He went in the other room. That's something he would never have done before. But you followed him.

Mother: I did, didn't I.

Me: So we have some more work to do. But we're getting there.

Finally, some children find the "thinking things through" part of dealing with frustration to be a major struggle. Basket B can be helpful in this regard because it's on Basket B issues that you have the opportunity to be clear about what problem needs to be resolved and provide suggestions, model effective problem solving and compromising, and help the child learn to think clearly. The ultimate goal, of course, is to help the child do the thinking through on his own.

Some children may also benefit from formal, systematic problem-solving training, in which they are taught how to think through problems in a planned sequence. Dr. Myrna Shure describes this approach in her excellent book *Raising a Thinking Child.* In such training, children are taught that they must first identify the problem at hand, think about all the possible solutions to the problem, then think about the likely outcome of each possible solution, select the solution that would seem to lead to the optimal outcome, and then monitor and evaluate the outcome.

But a lot of the children I work with aren't yet able to do this much processing on their own when they're frustrated. They often need additional help thinking things through, such as assisting them in generating alternative solutions or recalling adaptive ways in which they've handled similar situations in the past.

Child [*to younger brother*]: You dumb-ass. I hate you!

Parent: What's going on, fellas?

Child: He wrecked the Legos I was building! He always does this! He does it on purpose!

Parent: I know how frustrated you get when he does that.

Child: I hate him! I'm going to break something of his!

Parent: I think we need to think of something better for you to do than that, even though you're very frustrated right now.

Child: No! He wrecks my stuff, so I wreck his!

Parent: Hmm. I bet you can think of better things to do than that. Remember when you used to hit your brother when you were mad at him? You don't do that anymore.

Child: I don't care! He wrecks my stuff, so I wreck his!

Parent: We can think of something better than that. Let's think.

Child: I don't want to think. I get so mad when he does that.

Parent: I know you do. I'm sorry he wrecked your Legos. Let's think of what you did the last time he wrecked something of yours.

Child: I don't remember.

Parent: I do. You asked me to help you rebuild it. And your brother helped. It was fun.

Child: I don't want his help. I hate him.

Parent: You're still very mad at him. What else could you do right now so you're not so mad?

Child: I could go outside and play.

Parent: You could go outside and play if you wanted to. Anything else you'd like to do right now so you're not so mad?

Child: Uhm . . . I think I need to drink some juice. Then I want to rebuild my Legos. By myself!

Parent: That sounds easy enough. What kind of juice do you want?

Other such children—especially the more concrete ones—may initially need the skill of problem solving to be confined to a narrow set of common situations. In such instances, we can "concretize flexibility" by providing them with a simple framework of concrete solutions tied either to specific feelings or to the specific situations in which they're having the most difficulty. The mantra for one five year old with ADHD: *If I get frustrated, I need to tell Mom and ask her for a hug.* For a seven year old with ADHD and delays in expressive language: *If I start to get frustrated, I need to ask the teacher for help.* For a very anxious, disorganized nine year old: *When I start getting anxious about the weather, I need to find Mom or Dad so they can help me think about what can happen.* For a twelve year old diagnosed with bipolar disorder: *When I have a disagreement with someone and start to get heated up, I need to leave the situation and come back to the discussion later.* For an executively impaired thirteen year old: *When my friends start to tease me, here are some things I can say that might get them to stop.* Once a child has succeeded in dealing with frustration under circumscribed conditions, his framework of solutions can slowly and systematically be expanded to other situations. Of course, our ultimate goal is to move beyond rote solutions and predictable problems and to lay the foundation for resolving problems across multiple situations. The best way to do so is to point out how a problem confronting the child in one situation is similar to problems he's faced in other situations, so he'll begin to understand that although the specifics of a given problem may vary, there are similarities that cut across all problems to which his problem-solving framework can be applied.

Don't forget, your child has the potential to become overwhelmed by frustration at any point in the process. So no matter what part of the recognizing-expressing-thinking-through sequence you're trying to train, keeping your child coherent in the midst of frustration is still Job Number One. In other words, don't forget the necessity for continuously reading how your child is responding to your help. If he says—in whatever words he can muster—that he's becoming overwhelmed by the assistance you're giving, believe him. Take a break. Give him some time to collect himself. Try a new tack. Come back to it later.

Drama in Real Life
Getting There

"This was a horrendous week," said Danny's mother in our fourth or fifth session.

"It was?" Danny asked, a little surprised.

"Yes, it was!" said the mother emphatically. "I got called every name in the book this week—repeatedly."

"But I didn't hit you, and that's good, right?" said Danny.

"Look Danny, it's fine that you haven't hit me lately," said the mother, "but you're still attacking me with words."

"Like when?" asked Danny.

"Like this morning when I was trying to help you get ready for school on time and you were late for the car pool. You called me a 'bitch' because I told you the car pool was waiting for you in the driveway."

"I'm sorry." Danny slid over on the couch and tried to hug his mother.

"I don't want to be hugged right now." The mother tried to escape her son's grasp. "I'm sick of how I'm being spoken to."

"I won't do it anymore," Danny pleaded. Then he looked at me, tears welling up in his eyes. "Am I going to be like this forever?"

"I doubt it," I said. "But it's obvious we've got more work to do. Tell me"—I looked at the mother—"what did you do when Danny called you a bitch?"

"I got very angry. . . . I told him he couldn't speak to me that way . . . and if he was going to speak to me that way, I wasn't going to make him breakfast anymore," she replied.

"Then what happened?" I asked.

"He got more upset," the mother answered. "He screamed something about how the reason he swears so much is because of the way I've raised him."

"What happened next?" I asked.

"We started screaming at each other about how I've raised him," said the mother.

"But that has nothing to do with Danny being late for the car pool," I observed.

"Yes, that's true, but I'm not going to put up with that kind of talk from him," said the mother.

"You've been telling Danny you're not going to put up with that kind of talk from him for a very long time, yes?" I asked.

"Yes, of course I have," she replied.

"Has saying that helped him swear any less?" I asked.

The mother paused. "No."

"Has all the punishing you've done over the years in response to his swearing helped him swear any less?" I asked.

"Not that I've ever noticed," she said. "It just makes him scream even louder."

"Then it may be time to help him find better words to let you know he's frustrated," I said.

"What does that mean?" she asked.

"It means that, at the moment, Danny apparently isn't able to let you know he's frustrated in a way that isn't offensive to you," I said. "So, instead of being offended or threatening consequences that haven't worked, you'd probably be better off helping him find better words."

"Help him find better words? I've never heard of such a thing," she said. "This is not the universal approach to parenting."

"I'm not sure what you mean by 'universal,'" I said. "But whatever you mean, it's pretty clear that Danny hasn't responded to it very well."

"So what do I do when he starts screaming about how I've raised him?" the mother asked.

"I'm assuming that's just mental debris—the stuff that comes out of Danny's mouth when he's not thinking clearly. Don't forget, there's little to be gained by trying to process mental debris. Talking about it when Danny's frustrated certainly isn't going to achieve anything except to move him closer to a meltdown."

"It's not so easy to ignore him when he says that kind of thing," said the mother.

"I know," I said. "Initially, that's the hardest part. But once Danny figures out that you're able to get past his foul language and accusations and actually help him, I think he'll slowly begin communicating his frustration in a less offensive manner."

"What do you want me to do!" Danny insisted.

"You can try to let your mother know you're frustrated in a way that's less offensive to her," I said.

"I could try that," said Danny. "What should I say instead?"

"Well, what else could you say when you get frustrated?" I asked.

"I could say, 'I'm getting really frustrated!'" Danny suggested.

"That would be excellent," I said.

"What am I supposed to say if he swears at me?" the mother asked.

"Maybe something like, 'I can see you're getting very frustrated,'" I suggested.

"That's queer!" Danny interjected. "She'd sound too much like a psychologist."

"What would sound better to you?" I asked.

"Uhm . . . how 'bout 'Damn, you're frustrated!'" he volunteered.

I looked to the mother for her opinion of this suggestion.

"You know, given what usually comes out of my mouth, that would actually be a major improvement."

"Super," I said.

"So once we've determined he's frustrated, what do I do next?" the mother asked.

"First, let him know you're happy to help," I said. "Then the trick is for Danny to let you know what he's frustrated about."

"Oh, it's usually pretty obvious," said the mother.

"So you're never at a loss for what he's so upset about?" I asked.

"Not usually," said the mother.

"OK," I said, turning to Danny. "Danny, when you're frustrated, do you want your mother's help or do you want her to just leave you alone?"

"Sometimes I want her help and sometimes I want her to leave me alone," he said.

"Hmmm. Are you pretty clear about what you want when you're frustrated?" I asked.

"Sometimes . . . uhm . . . usually . . . I . . . I don't know," he sputtered.

"How good are you at figuring out what would be most helpful to Danny when he's frustrated?" I asked the mother.

The mother paused. "I don't know . . . I've been so caught up in the swearing that I never really considered the possibility that he might need my help."

"He does," I said. "Badly." I turned to Danny again. "Danny, you've got two challenging goals this week. First, I want you to try to let your mother know you're frustrated in a way that isn't offensive to her, like we just rehearsed. Second, try to let her know how she can be most helpful to you when you're frustrated. What do you think?"

"I'll try," he said. "I might slip a few times."

"Danny will not be perfect at this," I said to the mother. "He's going to need your help expressing his frustration in ways that are more appropriate. And if he's receptive to it, he probably could also use some help figuring out how to deal best with the problem that's frustrating him. If he wants to be left alone, leave him alone. What do you think?"

"This is very different for me," she said. "I'll try." She looked at Danny. "I might slip up a little, too."

"Well, I think this is a very good thing for you both to be trying hard at," I said. "And something that's going to take a little getting used to for both of you. Here's the good part: This is the beginning of you guys being able to talk to each other about frustrating things without attacking, swearing, hitting, or making each other feel bad. That's pretty exciting. And I think we're getting there."

Let's summarize what we've been up to in this chapter. While a user-friendlier environment can help set the stage for more adaptive parent-child interactions and a major reduction in meltdowns, and the baskets can help you quickly establish priorities and decide what issues are worth inducing and enduring meltdowns over, there are two components of a user-friendlier environment that often require closer consideration: identifying the specific situations that routinely and predictably cause your child significant frustration and addressing the specific factors that contribute to this inflexibility and explosiveness.

How you deal with these situations and factors depends on three important considerations:

- How important it is that your child successfully master the demands of the situation right now—in other words, whether it's a high priority.

- Whether your child is realistically capable of mastering the demands of the situation right now—in other words, whether he has the skills.

- If your child is not capable of mastering the demands of the situation, whether it's realistically possible to address the factors that are contributing to his difficulties right now so you can help him develop the skills that would make such mastery possible. In other words, could your child successfully meet the demands if he was given some extra help?

Your answers to these questions will help you decide what to do. Don't forget, there's a whole universe of "what to do's" out there, and "better motivating" is only one possibility. The trick is to find the "what to do's" that match your understanding of your child's needs. The suggestions presented in this chapter are far from exhaustive, so if your current universe of options isn't getting the job done, seek the guidance of a mental health professional whose universe is wider.

Instead of asking yourself, "What's it going to take to motivate this kid to behave differently?" you've begun to ask, "Why is this so hard for my child? What's getting in his way? How can I help?"

Just in case you hadn't noticed, something we laid the seeds for in Chapter 7 sprang to life in this chapter: Children are becoming active participants in their own treatment. They have begun to assume some ownership of their difficulties and to be involved in working toward solutions. These children can take their problems seriously, can develop at least a basic understanding of the factors underlying these problems, can discuss these problems, can participate in problem solving, and can come to see adults as potential allies—as long as the adults get past the motivational thing, create user-friendlier environments, and try to get a handle on the specific deficits in skills that are preventing these children from responding adaptively. Instead of asking yourself, "What's it going to take to motivate this kid to behave differently?" you've begun to ask, "Why is this so hard for my child? What's getting in his way? How can I help?"

10
All Things Considered

We've come a long way in nine chapters. We started by taking a close look at how we conceptualize and interpret inflexible-explosive behavior in children, with particular emphasis on how your explanations for this behavior will greatly influence the ways in which you respond to it and attempt to change it. I encouraged you to try a new explanation on for size, namely, that many of these children are not choosing to be explosive and noncompliant but are delayed in the process of developing the skills that are critical to being flexible and tolerating frustration or have difficulty applying these skills when they are most needed. These difficulties may emanate from a variety of physiologically based pathways, including difficult temperament; deficits in executive skills, social skills, and sensory integration; mood and anxiety disorders, and impaired language-processing and nonverbal skills. Coming to a clear understanding of the factors that fuel a child's difficulties is the first step in trying to figure out how to help the child overcome them.

We took a closer look at the way adults often respond to inflexible-explosive behavior. I suggested that many of the popular motivational strategies that adults tend to apply to such behavior may not achieve the desired effect; although motivational strategies make the possible more possible, they do not make the impossible possible. In other words, it makes little sense to try to motivate skills of which a child is presently incapable. Doing so has the potential to set the stage for, and then perpetuate, adversarial interactions, which is exactly what we wouldn't want to see happen. My firm belief is that many of these children are already motivated; if they could be more flexible and handle frustration more adaptively, they would. They're waiting for us to provide them with the type of help they need.

Next, we began thinking about what that help may actually be. We started by striving to clear the smoke through the creation of a user-friendlier environment, which includes having adults (1) all be on the same wavelength in understanding the child's difficulties; (2) be much more judicious about the demands for flexibility and tolerance of frustration being placed on the child; (3) identify—in advance—specific situations that may routinely lead to inflexible-explosive episodes; (4) get better at reading the warning signals and take quick action when these signals are present; (5) interpret behaviors that occur during vapor lock and meltdowns for what they really are; (6) understand the manner by which different adults may actually be fueling the child's inflexibility-explosiveness; (7) use a more accurate common language to describe various aspects of the child's inflexibility-explosiveness; and (8) come to a more realistic vision of who the child is and what he can become.

The baskets framework is intended to help you achieve several of these goals. Basket A—which contains behaviors that are high priorities and worth inducing and enduring meltdowns over—helps you maintain safety and your role as an authority figure. Basket B—which contains behaviors that are high priorities but not worth inducing and enduring meltdowns over—helps you teach your child how to resolve conflict in a way that promotes communication, problem solving, relationship building, and perspective taking. Basket C—which contains all the behaviors that aren't important anymore—helps you remove as many unnecessary frustrations as possible from the path of your child's cognitive wheelchair. The user-friendlier components and the baskets share an important common theme: intervening before the child is at his worst—before and during vapor lock and at the crossroads—rather than after he's already at his worst—during and after meltdowns.

By clearing the smoke, you've set the stage to help your child develop the thinking and communication skills needed to deal with frustration more adaptively. And you've read about some of the skills you may actually want to train your child in and some ideas about how to do it. We also took a close look at sibling issues and family communication patterns and how they may best be addressed.

Though we've come a long way, you probably have some questions that haven't yet been answered. This chapter is devoted to answering some of the questions I hear most often.

Question: "Isn't this a very passive approach to parenting?"

No . . . in fact, this is a very active approach to parenting. Remember, there is no easy way to parent an inflexible-explosive child. Hard work is a given. Let's just make sure you have something to show for it.

Question: "How much progress should I expect from my child, and how fast? When should I stop doing the whole user-friendlier environment and baskets thing?"

Children and parents vary widely in terms of how quickly they respond to this approach. The first goal is to take the fuel out of the fire as quickly as possible; in other words, for there to be a fairly immediate decrease in unsafe behaviors and a dramatic shift in the way parents respond to and communicate with their children. Some families are able to achieve this goal in a few weeks, others take several months, and yet others take longer still. Some children continue to have occasional, residual unsafe episodes for a few months, but such episodes are often far less intense and fizzle out a lot faster.

As you also know, once the smoke clears—and parents and children are communicating more adaptively and starting to feel better about each other—the stage is set for work on the skills that will help the child handle frustration more adaptively. The timing of this process varies from child to child and family to family as well. Dealing with frustration is a continuous learning process even for those of us who aren't children anymore. So we're basically looking for slow, steady progress in the child's capacity for empathy and perspective taking; improvements in the child's ability to anticipate, recognize, express, and think through solutions to frustrating situations independently; a longer fuse; less frequent vapor lock; and fewer meltdowns. Reductions in swearing and lying often take longer, but I still expect slow, steady progress in these domains, too.

Many parents want to know when they can get back to their old way of doing things. My question is, Why would you want to? A lot of parents begin to use the approach described in this book thinking that eventually they'll be able to put lots of things into Basket A. In reality, as parents and children get better at communicating, negotiating, and compromising; as parents get better at creating a user-friendlier environment and implementing the baskets; and as the relationship between parent and child steadily improves, the importance of Basket A actually *diminishes*. Many families I've been working with for three or four months hardly ever mention Basket A anymore.

Question: "So does using this approach mean that rewards and punishments are completely out of the picture?"

Not necessarily. But my hope is that by now, you have a realistic sense of what the motivational approach can and can't help you achieve and an awareness of the special care required when using rewards and punishments with an inflexible-explosive child. The real question is this: Will additional motivation enhance your child's performance at any point along the way? The answer: maybe.

The first thing I'd want to be convinced of is that the child would actually benefit from additional motivation. If so, consequences are still in the running. If not, hold off.

Second, I'd want to be sure that the motivational strategies are worth their potential price. My rule of thumb is that motivational strategies should be applied only to behaviors that are in Basket A. Since you're willing to induce and endure meltdowns only over behaviors that are in Basket A—and meltdowns are likely to occur if your child loses or fails to obtain a privilege or is punished—you should use motivational strategies just for Basket A behaviors. For example, a few of the children I've worked with have earned and lost privileges contingent upon having a safe day. Was I hopeful that a little added reminding and motivating would be helpful? Yes. Did I think incentives would be the primary reason the child would start behaving in a safer manner? No, I thought the user-friendlier environment and baskets would be the primary reasons the child would start behaving more safely.

Third, I'd want parents to be certain that they're willing and able to enforce whatever consequences are being applied. If so, keep going. If not, do not pass go. Consequences that can't be enforced aren't consequences at all; they just detract from your credibility, which is something we're trying to avoid.

Finally, I'd want you to have faith that a consequence administered on the back end of one inflexible-explosive episode is going to be accessible and meaningful to your child the next time he's frustrated. In other words, I'd want you to be confident that the next time your child enters vapor lock, he could navigate his way through the following thought process: "The last time I got frustrated I broke a lamp and my mom grounded me for a week. I don't want that to happen again. So I'd better think of a better way to handle my frustration this time. Let me think of what the possibilities are." If you're not confident of your child's ability to benefit from consequences in this manner, then consequences may be of limited benefit.

Mother: So I'm not supposed to punish my kid anymore?

Me: Tell me about the last time you punished him.

Mother: The other day I told him to turn off the TV and do his homework. He called me a bitch, so I took away TV for three days.

Me: Is that a punishment you're able to enforce?

Mother: Enforce? Not really . . . in fact, he was watching TV the very next day.

Me: What did you do?

Mother: What could I do? There really aren't any punishments I can enforce anymore.

Me: I guess you shouldn't be punishing much then.

Mother: I shouldn't be punishing much? Then what should I do?

Me: I think if we make some fundamental changes in the way you two interact, he'll end up swearing at you a lot less and you won't have much cause to punish him anymore. By leading with punishment—in other words, by having punishment be your primary way of responding to behavior you don't like—you end up cutting off communication the instant there's a disagreement or something happens that you don't like. You also end up making yourself the target of his frustration, rather than whatever it was that was frustrating him in the first place. There's not much sense in putting yourself in that position. And since you can't enforce your punishments anyway, you end up feeling ineffective and disempowered.

Once communication between the parents and children improves, I sometimes encourage the parents to involve the children in discussing whether a punishment is warranted for a certain inappropriate behavior and what that punishment should be. Children are often much harder on themselves than adults would be, so I don't usually worry about whether the punishment they come up with will be stiff enough. Enforcement of agreed-upon punishments can be a problem: Sometimes children agree that a certain punishment is appropriate (for example, an earlier bedtime) and then melt down when the punishment is invoked. This may be a sign that they aren't quite ready for the delay between agreeing on a consequence and following through with it, and it brings us back to the original question of whether the punishment truly facilitates learning that makes the meltdown worth inducing. I'm also enthusiastic about engaging children in discussions about how they can "make amends" for an act committed in the midst of frustration. Such discussions should not take place during or right after meltdowns but rather when coherence has been fully restored. Here's what such a discussion might sound like:

Parent: Carlos, we need to talk a little about the table you broke yesterday.

Child: I said I was sorry.

Parent: I know, and it was very nice of you to say that. But I still feel very bad about the broken table, and we still need to figure out what you can do to help me feel better.

Child: Like what?

Parent: I don't know. Can you think of anything you could do that would make me feel better?

Child: You could punish me.

Parent: I don't know if punishing you would make me feel better. I was thinking there might be some things you could help me with around the house.

Child: I could sweep the floors.

Parent: There's an idea. That would be very helpful. Is that something you'd do to help me feel better?

Child: Yes. Or I could help you take out the trash.

Parent: I think it would be most helpful for you to sweep the floors. That would make me feel a lot better. Maybe next time you get frustrated you could let me help you instead of breaking the table.

Child: I'll try.

Whether it's a punishment or making amends, children are often capable of coming up with outstanding ideas, which helps their parents become "solution facilitators," rather than "solution imposers."

Question: "What about time-out?"

Some children find time-out to be a good place to calm down when they're frustrated; most of the ones I work with don't. Indeed, the meltdowns of many of the children I work with are actually exacerbated—sometimes dramatically so—if someone makes any kind of physical contact with them when they're frustrated. So if your child tends to calm down in time-out and will go to and stay in time-out without having to be restrained and without destroying property, go for it. If time-out simply fuels your child's meltdowns, forget it. Even under optimal circumstances, time-out is typically not recommended for older children and adolescents.

On the other hand, I've often found it productive to help parents and children agree to go their separate ways—with each going to different, designated rooms of the house—when it becomes obvious that a disagreement or discussion is going poorly. Not all inflexible-explosive children will follow through on this plan, but a surprising number will. The discussion resumes after everyone has regained sufficient coherence.

Question: "But I still have the feeling that some of my child's behavior is planned and willful. What now?"

The last thing I'd want you to try to do at times when your child is in vapor lock or at the crossroads is quickly try to figure out whether his behavior is planned or unplanned. You don't have a whole lot of time to spare. So there are essentially two possible mistakes you can make at such moments. The first is to think your child's behavior is unplanned and unintentional when it really isn't. The second is to think your child's behavior is planned and intentional when it really isn't. If you have to make an error, I'd rather see you make the first one. In other words, when in doubt, respond as if your child's behavior is unplanned and unintentional. I don't think you have all that much to lose. The ramifications of the second error are much more serious.

Question: "When can I add more behaviors to Basket A? What do I add first?"

As you know, safety is always in Basket A; going to school and getting there on time are usually in Basket A as well. If your child is already mastering these behaviors—the vast majority eventually do—and you're confident that he's capable of handling a little more, you're probably ready to add one or two high-priority behaviors to Basket A. Just don't forget the Basket A litmus test: These must be behaviors your child is capable of performing on a fairly consistent basis and ones you're actually willing and able to enforce. Also try to remember that Basket A isn't the most important basket.

One of the behaviors that parents are usually eager to put in Basket A as soon as possible is swearing. I usually discourage them from doing so until we've helped their child develop more adaptive ways of expressing frustration. As you now know, this goal usually involves trying really hard not to take the swearing personally and to interpret foul language that occurs in the context of frustration as an indication that the child is stuck. Which means you'd probably be helping your child most by—I don't know any other way to put it—helping unstick him.

Question: "OK, I get the idea about swearing. What about lying? Basket A?"

A while ago I heard an interesting idea about why children with ADHD tell lies. Imagine being a child who has a whole bunch of knowledge stored in his brain about appropriate ways of behaving. Also imagine being so impulsive that this store of knowledge seldom prevents you from behaving inappropriately. In other words, you're continuously behaving in ways you "know" are inappropriate, but you aren't gaining access to the knowledge that your behavior is inappropriate rapidly enough to keep you out of trouble. It's like there's two of you—the you who engages in inappropriate behavior and the you who clearly knows better. The stage is now set for lying.

Here's how. You behave inappropriately. Then, after a brief delay, your knowledge about appropriate behavior finally kicks in, and you find yourself in the position of having to explain why you just behaved in a way you already "knew" was inappropriate. Most children with ADHD don't have the presence of mind to say, "I know I shouldn't have done that; I just didn't access the information in time." What many do instead is fall back on one of the most primitive defense mechanisms: denial. So what comes out of their mouths is "I didn't do it." This pattern can lead to some fascinating conversations:

Parent: Peter, why did you just push your sister?

Peter: I didn't.

Parent [*incredulous*]: Peter, I just watched you push your sister.

Peter: No you didn't.

Parent: Peter, what are you talking about? You're lying!

Peter: No, I'm not.

Is this a manipulative, lying child or a child who is struggling to comprehend why he does things he knows he shouldn't?

Other children with ADHD are not good at reflecting long enough to come up with accurate information, but they are good at quickly figuring out what adults want to hear. The resulting "lie" sounds something like this:

Mother [*entering her daughter's bedroom*]: Roberta, is your homework done?

Roberta [*whose mind was not on homework*]: Yes.

Mother [*scanning Roberta's homework assignment sheet*]: All of it?

Roberta [*still barely accessing the "Did I do my homework?" file in her brain*]: Yes.

Mother: Roberta, there's a good hour's worth of homework here. You haven't even been home an hour.

Roberta [*unable to say, "Gee, mom, I guess I wasn't able to access my 'Did I do my homework?' file as quickly as my 'What does my mother want to hear?' file and my 'What will my mother do if she knows I've now lied?' file*]: Oh, I did it all at school.

Mother: When did you have a chance to do it all at school?

Roberta: Oh, Mr. Stearns always gives us time to do our homework.

Mother: Mr. Stearns gives you an hour to do your homework?

Roberta: Yep.

Mother: You're telling me the truth?

Roberta: Yep.

Since lying generally does not involve safety and is often a highly impulsive, unplanned act, my tendency is to put it in Basket B. I'd much rather deal with lying in the context of a discussion about parent-child trust than simply toss it into Basket A.

As long as we're on the topic of impulsivity, a brief remark about the word *manipulative,* which I've found to be a remarkably overused, overrated explanation for the behavior of inflexible-explosive children. To me, the act of manipulating requires a fair amount of forethought, planning, affective modulation, and calculation—qualities that are in short supply in the vast majority of the inflexible-explosive children I know. Given that few of us enjoy being manipulated, believing that a child is being manipulative often causes adults to behave in counterproductive ways and hinders their consideration of more accurate explanations.

And as long as we're on the topic of overused terminology, I've always been curious about what adults are really up to when they tell a child he has to "take responsibility" for his actions. I don't put a lot of stock in having children "take responsibility," since I think of the phrase as being fairly meaningless. I'm certainly not interested in having adults and children fight over it. If a child is taking his difficulties seriously and actively trying to change—even if he has trouble saying "OK, I did it"—I'm happy.

Question: "I don't always have time to compromise—like when my child is getting ready for school or when it's time to go to bed at night. What then?"

Time pressure is one of the great fuelers of frustration in both parents and children and most commonly comes into play during morning and evening routines. Unfortunately, training a child who's underaroused in the morning or overaroused at night to be more sensitive to issues of time is extremely difficult. Once again, your best hope during these times of the day is to make adaptations—often by judiciously establishing priorities and removing as many frustrations as possible—to make these things more manageable for your child. Your other option is to increase your involvement during these times of the day, not by badgering but by performing some tasks for your child that he seems unable to perform in a timely way by himself right now. Don't forget: It takes a lot less time to help your child get moving than it does to deal with a meltdown.

Question: "My child becomes frustrated about things that don't involve interactions with me or other people. He just gets really frustrated with something he's doing, like playing Nintendo, or about something that makes no sense. Or sometimes he has a delayed response to a frustration that happened earlier in the day. The baskets don't work so well then because there's nothing to compromise on, but he stays upset and it affects the tone of our interactions. What then?"

It's true, there are times when there's nothing to compromise about because your child's frustration didn't involve you or another person. And there are instances in which a child's frustration at the moment is a delayed response to an earlier frustration, such as something that happened in school. Such delayed responses are probably even more difficult for a child to articulate. But your role remains the same—to find out what's frustrating your child by helping him stay coherent enough to tell you what the problem is or by suggesting plausible hypotheses and helping him figure it out—and then showing him that you can help him think things through well enough to come to a resolution. Eventually—when he's able—the goal is for him to figure it out and resolve it on his own.

Mother [*standing on the front walk to the house*]: Charlotte, we're waiting for you to get in the car so we can go to the beach.

Charlotte [*standing in the front doorway—vapor lock*]: I'm not going.

Mother: What? Charlotte, you love the beach.

Charlotte [*backing into the house—crossroads*]: I said I'm not going!

Mother [*moving toward the front door*]: Charlotte, your brother and father are already in the car, we're in a hurry, and I don't feel like going through this with you right now! Go get me my keys and let's go!

Charlotte [*slamming and locking the front door to the house—meltdown*]: Stay away from me! I'm not going!

Mother [*outside the locked front door*]: Charlotte, you unlock this door right now! [*Turning to her husband in the car*]: Honey, do you have your keys?

Husband: No. Why?

Mother [*pulse pounding, turning back to the locked front door*]: Charlotte, open the door goddammit! This isn't funny.

No response from inside the house.

Father [*arriving at the front door*]: What's going on?

Mother [*through gritted teeth*]: Your daughter has informed me that she's not going to the beach and has locked us out of our house.

Hmmm. That looks like an example of how *not* to do it. Now, through the magic of book writing, let's do something that you don't have the luxury of doing in real life: rewind the tape and try a user-friendlier tack.

Charlotte [*vapor lock*]: I'm not going.

Mother: You're not going?

Charlotte [*crossroads*]: No, I'm not going!

Mother [*recognizing vapor lock but at an absolute loss for what might be frustrating her daughter*]: What's the matter, honey?

Charlotte [*rubbing her eyes*]: I just don't want to go.

Mother [*squatting down to Charlotte's level*]: Is there something about going to the beach today that's frustrating you? You usually can't wait to go to the beach.

Charlotte: It's too early to go to the beach!

Mother [*still at a loss*]: I don't know what you mean that it's too early to go to the beach.

Charlotte: We don't usually go to the beach until after church. We never go to the beach in the morning.

Mother [*hiding her incredulity*]: It's bothering you that we don't usually go to the beach in the morning?

Charlotte: We never go to the beach in the morning. We can't go right now.

Mother [*lightbulb now on*]: I'm glad you told me what was the matter. Let's think about this a little. Maybe we can figure out what to do.

No meltdown. Much nicer.

Remember, if your child is routinely becoming frustrated over the same things, you've got three options: help him avoid the situation, make alterations or adjustments so the situation is more manageable, or train him in compensatory skills.

Question: "My child has to have each compromise spelled out in perfect detail. Is this reasonable?"

Many inflexible children don't handle ambiguity in their lives well, and this extends to their compromises. So statements like, "OK, we'll do that later" or "We'll go there soon" or "You can do that for a while" have the potential to fuel their frustration, even in the context of compromising. For such children, clarity is your friend and ambiguity your enemy.

Trent [*sitting in the backseat of the family car—vapor lock*]: I need something to eat.

Mother [*perking up in the passenger seat and immediately recalling the potentially unpleasant ramifications of prolonged hunger in her son but failing to remember the dangers of ambiguous courses of action*]: We'll stop for something very soon.

Trent [*vapor lock deferred*]: OK.

Five minutes of silence elapse.

Trent [*with agitation—vapor lock déjà vu*]: I thought you said we were stopping to eat!

Mother: I said we'd stop soon.

Trent [*with significantly greater agitation—advanced vapor lock*]: I can't wait! You said we were stopping!

Father [*from the driver's seat*]: Your mother said we'd stop soon; now put a lid on it!

Trent [*very loudly, kicking the back of his father's seat—crossroads*]: You guys are such fucking liars! You always do this! You say you'll do something and then you don't!

Mother: Look, we'll stop for food as soon as we can.

Kaboom.

On a related topic, some children don't adapt well when a compromise doesn't go exactly as planned. Mike, a remarkably rigid thirteen year old, had negotiated a compromise with his mother over when (12 noon on Saturday) and how (with her help) he'd clean his room. Mike was actually eager to get his room straightened up, but lacked the organizational skills to do it on his on. Unfortunately, the mother was delayed by another commitment and wasn't around at 12 noon to help Mike clean his room. This change in plan proved to be a major obstacle for Mike's cognitive wheelchair. When the mother arrived home at 1:30 P.M. and suggested that they begin cleaning the room, Mike entered vapor lock. At 1:31, the mother insisted that they clean the room. Mike's agitation increased. The mother insisted further. At 1:32, Mike melted down. What Mike needed was a complete renegotiation of the original compromise. He was just that rigid.

Some children can handle greater ambiguity. You have to figure out what your child can handle.

Question: "Is language therapy helpful for inflexible-explosive children? How about occupational therapy?"

For children with linguistic impairments, language therapists can be very helpful at facilitating many of the skills I've been discussing throughout the book; specifically, helping children label their emotions, articulate their frustrations, and think through solutions. I've also seen a talented occupational or physical therapist be very helpful. Such therapists can help children feel a greater sense of mastery over the barrage of sensory stimuli coming at them from the world, thereby reducing their cumulative levels of frustration. Other similar types of therapies can be useful as well. I was recently sitting in a meeting at a school discussing one of the children I was working with and his adapted physical education teacher—a wonderful man named John Passarini, who works in the Wayland, Massachusetts, Public Schools—was describing his work with the boy. He was talking about how far the boy had come in learning how to shoot a basketball, but that he had to be taught in very slow, very gradual steps. John was commenting that this gradual training of skills had greatly reduced the boy's frustration during basketball games, and that he had become a more active participant in other sports as well. I remember thinking, "My goodness, this man is doing exactly what I try to do with these kids!" Clearly, there's more than one way to skin a cat.

Question: "Many of the examples you've given relate to younger children. My inflexible-explosive child is fifteen. Any special suggestions?"

My emphasis on younger children reflects my bias that the best time to help an inflexible-explosive child is well before his level of alienation reaches its peak, he's begun affiliating with similarly alienated peers, and he's become much harder to reach. Ultimately, the language we'd use to help an inflexible-explosive child would probably be more sophisticated for a fifteen-year-old than for a four-year-old, but the emphasis on clearing the smoke and training the skills necessary for improved flexibility and tolerance for frustration would be the same. The precise skills we'd try to train would, of course, depend on each individual child no matter what the age.

Older kids may engage in a variety of behaviors that may make basket thinking more of a challenge. For example, parents of many teenagers tend to be very invested in placing drug use and sexual promiscuity in Basket A. While it's absolutely true that these behaviors are unsafe, other components of the Basket A litmus test may not be satisfied, especially as related to enforcing abstinence. So while there are some circumstances in which such behaviors end up in Basket A, it's more typical for them to be placed in Basket B, where parents and teenagers can start talking to each other about them. In other words, I have more faith that rebuilding your relationship with your teenager will reduce dangerous behavior and keep your child from being arrested than pretending you can enforce the unenforceable.

Question: "Things were going really well, and then the bottom fell out. What happened?"

Sometimes a minor change can throw a monkey wrench into the works. Sometimes something that happened outside the context of your interactions with your child, for instance, at school, has the same derailing effect. Sometimes medications need to be adjusted. Your goal is to try to get a handle on what's going on so you can help your child get back on track.

Recently I received a phone call from the mother of a very inflexible six-year-old girl I'd worked with about six months before. It was just after New Year's and the mother was beside herself.

"Monica has completely deteriorated," she reported with great distress. "This was an absolutely horrendous Christmas."

"What do you think is going on?" I asked.

"I have no idea," said the mother. "All she could tell us was, 'The Christmas tree is in the wrong place.'"

"What does that mean?" I asked.

"I don't know," said the mother. "Although this was our first Christmas in our new home."

"Is that so," I said. "And things were going fine before Christmas?"

"Never better," said the mother.

"And how are things now that she's back at school?"

"She's been great the past few days."

"Was she sick during Christmas?"

"Perfectly healthy," said the mother.

Monica wasn't on medication, so I didn't entertain the notion that her psychopharmacology might need adjusting. And since she'd been fine before and after Christmas, I assumed there wasn't a seasonal component to her sudden difficulties.

"Well, then, my best bet is that spending Christmas in a new place threw her for a loop," I said. "I guess we haven't yet experienced the full repertoire of things she's capable of being inflexible about. And being home from school took her out of her usual routine."

"You know, now that you say that, I think you may be right," said the mother. "Who would've thought that spending Christmas in a new house would throw her like that?"

Question: "This whole thing isn't working too well. What's going on?"

In my experience, one of the most likely explanations is that you're neglecting Basket B. Many parents buy into the philosophy of a user-friendlier environment but don't make the fundamental changes in how they interact with their child that such an environment is intended to facilitate. What they do instead is continue to react as if everything is still in Basket A and fail even to consider whether a situation really belongs in Basket B or C. Then, when their child reacts strongly to a request or change in plans, the parents default into Basket C and end up feeling like they're still giving in. They're back to responding on the back end. Because Basket B is being neglected, there's little negotiating, compromising, communicating, or relationship repairing going on. In reality, this pattern differs little from the one that brought them into treatment in the first place. Nothing has really changed.

Child: Remember the allowance money you put into my account when I went to camp? Well, I didn't spend all of it. Can I have the leftover money?

Father: Good luck.

Child: Why? You put the allowance I usually get every

month into my camp account. I didn't spend it all. Now I want the leftover money.

Father: Sorry, Charlie. No spendy, no keepy.

Child: Why not! That's not fair! I didn't spend it all!

Father: I don't care. That money was for camp. You didn't spend it. That's your problem, not mine.

Child: Goddammit! You're such a fucking jerk! I want my fucking allowance money!

Father: Talk to me like that, and you'll lose your allowance for this month, too.

Kaboom.

Later:

Mother [*exasperated, to father*]: Was not giving him the leftover allowance really in Basket A?

Father: What do you mean?

Mother: Well, I assumed by your handling of the issue that not giving him the allowance was in Basket A.

Father: Why did you assume that?

Mother: For starters, you never mentioned the word *compromise,* and you seemed perfectly willing to induce and endure a meltdown over it, which you did. It looked like Basket A to me.

Father: No, I was actually willing to negotiate over the leftover money.

Mother: You mean, the issue was in Basket B?

Father: Sure, I don't really care that much about the allowance money. I'm just not willing to compromise when he gets upset like that.

Back to Chapter 6. If your knee-jerk response to every situation looks like Basket A, you're going to induce and endure a lot more meltdowns than you intended. Basket B issues—those that are important or undesirable but over which you're not willing to induce or endure a meltdown—have to look, sound, taste, and smell like Basket B issues.

It's also possible that you've yet to fully revise your vision of who your child is. In other words, you're still determined to create a child who's able to respond to requests and changes in plan with calm compliance and adaptability. Back to Chapter 3: You may need help from a good mental health professional with this one.

Possibility number two: The stage has been set for more adaptive interactions within the family, but things are going really poorly at school and this situation carries over to the family. I'll discuss this issue in the next chapter.

Possibility number three: Your child is so irritable, agitated, and moody, or so hyperactive and impulsive, or has such a short fuse, or is so out of sorts so often that he can't benefit fully from the strategies you've been reading about. If you've done a good job of executing these strategies and you're still enduring countless meltdowns, it may be time to consider medical intervention. Although no book can substitute for competent psychopharmacology, an overview of medication and related issues is provided in Chapter 12.

11
User-Friendlier Schools

As hard as it is to help an inflexible-explosive child within the confines of your home, I think it may be even harder in a classroom. Can you imagine a teacher trying to implement the strategies described in the preceding chapters while simultaneously being responsible for teaching twenty-five to thirty other students, many of whom have other types of special needs? Like parents, most regular education teachers have never been responsible for changing the behavior of an inflexible-explosive child and have never received any specialized training to prepare them for this task (I assure you, my approach to dealing with my first inflexible-explosive child was a lot different than my current approach). How does the teacher keep his or her other students safe and under control while the inflexible-explosive student is falling apart over, say, a spelling test? What can other school staff—the principal, guidance counselor, school psychologist, and so forth—do to help? Does the child need an aide? Should he be in a special education classroom? A special school?

Fortunately, for a lot of inflexible-explosive children, these questions never need to be answered. For reasons that are not entirely clear, many of these children and adolescents with whom I work do *not* evidence serious signs of vapor lock or meltdowns at school. I've got a few theories about this phenomenon:

- *The embarrassment theory*: They'd be embarrassed if they melted down in front of their friends; they've been melting down in front of their parents practically since the word go, so it's not a big deal. Or the child feels safer melting down in front of his parents.

- *The tightly wound theory*: They've put so much energy into holding it together at school that they become unraveled the minute they get home, fueled further by normal late-afternoon fatigue and hunger.

- *The herd-mentality theory*: Because the school day tends to be relatively structured and predictable, it can actually be user-friendlier than unstructured downtime at home. For instance, if the child becomes confused about where he's supposed to be or what he's supposed to be doing while he's at school, he need look no further than his classmates for cues.

- *The chemical theory*: Teachers and peers often are the primary beneficiaries of pharmacotherapy, but the medications may have worn off by late afternoon or early evening.

I'm certain there are other possibilities. Of course, just because a child isn't melting down at school doesn't mean that school isn't contributing to meltdowns that occur elsewhere. Lots of things can happen at school—being teased by other children, feeling socially isolated or rejected, feeling frustrated and embarrassed over struggles on certain academic tasks, being misunderstood by the teacher—to fuel meltdowns at home. And homework can extend academic frustrations well after the bell rings at the end of the school day. So schools aren't off the hook for helping to achieve a globally user-friendly environment even if they don't see the child at his worst.

Nonetheless, some children actually *do* melt down in one form or another at school. You may recall that Casey, one of the children you read about in Chapter 4, had a pattern of running out of the classroom when he became frustrated by a challenging academic task or difficult interaction with a peer. When he wasn't running out of the room, he was melting down in the room, turning red, crying, screaming, crumpling paper, breaking pencils, falling on the floor, and refusing to work. Needless to say, because his low frustration tolerance caused him to shut down quickly on even mildly challenging tasks, there were concerns about whether his learning was suffering.

Danny, another of the children you've read about, was also capable of the occasional meltdown at school. On one particularly memorable day, the teacher designated him to hand out doughnuts to his classmates after recess. Following recess he hurried back to the classroom to hand out the doughnuts. But a parent-aide was already in the room and insisted on being the doughnut distributor. Danny attempted to explain to the parent that he had been assigned the task of giving out the doughnuts, but the parent would not be deterred. The shift in cognitive set demanded by this scenario was more than Danny could handle. Kaboom.

Teachers and schools have little choice but to think about

how to handle the Caseys and Dannys. We live in the age of the inclusion movement, which I endorse wholeheartedly. This movement, which has affected certain parts of the country more than others, has encouraged the practice of providing additional services for children with special behavioral and educational needs inside the mainstream classroom, thereby reducing—if it's done right—the stigma of having a different style of learning or behaving. Thus, the typical public school classroom now has multiple special-needs students, some of whom have disorders that teachers have never even heard of, let alone worked with, before. Teachers must therefore have knowledge not only of the curriculum, but also of the different emotional and behavioral issues presented by some of their students. Unfortunately, in many instances, teachers feel—justifiably—that they are not being provided with the kind of support or additional training they need to function effectively with their special-needs students and within the altered composition of their classrooms.

It is against this backdrop that teachers are asked to help inflexible-explosive children overcome their difficulties. And school personnel are as prone to back-end interventions as are any other adults, especially in relation to dealing with children whose frustrations cause them to disrupt the classroom process, behave disrespectfully toward adults, and become physically or verbally aggressive. School-initiated, back-end interventions have the same potential for creating an adversarial process and alienating children as do many parent-initiated interventions. I receive many phone calls from school administrators telling me that their disciplinary practices aren't working very well for the students to whom these practices are most frequently being applied, and asking for help. Thus, the need for a Plan B seems evident to me, for we have little to show for all the consequences—detentions, suspensions, expulsions, and so forth—that have been administered to the inflexible-explosive students who have passed through our schools.

I'm often astounded to hear that the primary reason such interventions are invoked is *for the benefit of the other students*. The rationale goes something like this: "We have to set an example for all our students; even if suspension doesn't help Lamar [whom you'll meet later in this chapter], at least it sets an example for our other students. We need to let them know that we take safety seriously at our school." This rationale begs the following question: "What message do you give the other students at your school if you continue to apply interventions that aren't actually helping Lamar behave more adaptively?" My answer: "That you're actually not sure how to help Lamar behave more adaptively."

Question: "What's the likelihood that the students who aren't inflexible-explosive will become inflexible-explosive were it not for the example we're making of Lamar?"

Answer: "As a general rule, slim and none."

Question: "What message do we give Lamar if we continue to apply motivational interventions to a pattern of behavior that isn't motivational?"

Answer: "We don't understand you, and we can't help you."

Question: "Under what circumstances do we have better odds of helping Lamar learn and practice better ways of dealing with his inflexibility and low frustration tolerance: in school or suspended from school?"

Answer: "In school."

Question: "Why do many schools continue to apply reactive, back-end interventions that aren't working for their inflexible-explosive students?"

Answer: "They aren't sure what else to do."

Question: "What happens to students to whom these interventions are counterproductively applied for many years?"

Answer: "They become more alienated and fall further outside the social fabric of the school."

Question: "Isn't this the parents' job?"

Answer: "Helping a child deal more adaptively with frustration does not fall exclusively under the parents' purview. In fact, it's the children whose parents are struggling—with poverty, psychopathology, unemployment, and so on—who may need extra help from schools the most."

Question [slightly off-topic]: "In our society, what is the ultimate back-end intervention, and how well does it train flexibility and the tolerance of frustration?"

Answer: "Jail. Very poorly."

Question: "Is there any way in which the other students may actually be able to help Lamar with his inflexibility and explosiveness?"

Answer: "I'm glad you asked!"

Time for Plan B.

Let's take the same principles that we've been applying to inflexibility-explosiveness at home and see how they might look at school. And let's begin with some of our main themes:

- Flexibility and frustration tolerance are *skills*. Some students are predisposed to have difficulty responding to the world in an adaptable, flexible manner. Thus, these students can be exceptionally challenging and frustrating to teach.

- Because inept parenting, poor motivation, attention seeking, and lack of appreciation for who's boss may not be the best explanations for these students' difficulties, standard teaching and motivational practices may be mismatched to their needs. These students may require a different approach.

- How we perceive and understand these students' inflexibility-explosiveness is directly related to how we'll ultimately respond to it (*"Your explanation guides your intervention"*). If we respond to the student in an inflexible, angry manner, then we will increase the likelihood of vapor lock and meltdowns, facilitate adversarial teacher-student interactions, and reduce the likelihood of improving the student's flexibility.

- I am more optimistic about a teacher's ability to respond differently to a student's inflexibility than I am about the student's ability to improve his capacity to be flexible, at least initially.

- If we create a user-friendlier school environment for an inflexible-explosive student and concentrate more on the front end than the back end, interactions between school staff and the student should improve. Once things begin to improve a little, the student should become more receptive to, and have a greater capacity to benefit from, more direct methods for improving his flexibility and tolerance of frustration.

Next, let's apply the components of a user-friendlier environment described in Chapter 6 to the classroom:

A user-friendlier school environment is one in which *all* the adults who interact with the child have a clear understanding of his unique difficulties. Before and when an inflexible-explosive child is stuck in the red haze of inflexibility, lots of adults in a school can serve as potential helpers. And lots of adults can take on the role of enemy by failing to take into account the student's limitations in the domains of flexibility and frustration tolerance. Parents are not the only ones who can act as temporary guides to steer an inflexible-explosive child around, or help him think through, frustrations; teachers can step up to the plate as well. Coming to a common view of a child and making sure everyone is on the same wavelength requires a lot of teamwork within the school environment. It's important for all the adults who are likely to interact with the student to be brought into the mix.

Parent-teacher collaboration is absolutely essential. When things are going poorly at school, parents have a tendency to blame the school staff, and the school staff have a tendency to blame the parents. This blaming misses the important point: The child is frustrating *all* of us, and *none* of us has done an incredible job of helping him yet, so let's see if we can put our heads together and come up with a plan that incorporates the best we all have to offer.

Some teachers are eager for a quick fix. But there is no quick fix for an inflexible-explosive student. And we can't jump into intervention until we have an answer to the question, "What's this kid's deal?" Restated, we have to put the hard work into looking closely at the pathways that seem to be associated with a student's inflexibility and low frustration tolerance before we'll even have a clue about what interventions may make the most sense. Any attempt to intervene before being able to answer "What's this kid's deal?" is akin to firing a shotgun randomly into the air and hoping to hit something good. And reaching a clear understanding of a child begins with competent assessment.

When assessments are performed in schools, they are often initiated by the classroom teacher because he or she has some concerns about the child's functioning in the classroom. As you've already read, a diagnosis is not the primary goal of an assessment, although it may be an important by-product. Nor should determining whether a child qualifies for special education services be the main thrust. Once again, the primary goal of an assessment is to achieve the fullest possible understanding of the child and the school environment in which he functions. Even if no learning disabilities are uncovered and no diagnoses rendered, the assessment process is not complete.

We still have a student whose teacher has serious concerns about his functioning! We still need to understand the factors underlying his difficulties in the classroom! Once we truly understand, logical interventions often become evident.

In a user-friendlier school environment, the adults try to reduce the overall demands for flexibility and frustration tolerance being placed on the inflexible-explosive student. In other words, there has to be a Basket B and C in the classroom. I'm told by my educator colleagues that good teachers are able to maintain a balance between the common needs of all their students and the specific needs of individual students. In other words, good teachers do a good job of teaching individual students. I've seen that it is possible to reduce demands for some students without stigmatizing them and without alienating others. One way to do so is to create a "responsive" or "prosocial" classroom, which you'll read about later in this chapter.

A user-friendlier school environment is one in which adults try to identify—*in advance*—specific situations that may routinely lead to inflexible-explosive episodes. This theme is critical to your attempts to intervene proactively, on the front end. As you've read, it's important for the adults who interact with the student at school to take a close look at the specific situations—including content areas, methods of producing work, contexts (large- and small-group discussion and instruction, independent seat work, working with a partner, story and circle time, recess, and lunch)—that may cause significant frustration. Then the same three questions must be considered:

- How important is it that the student successfully master the demands of the situation right now? In other words, is it a high priority?

- If it is a high priority, is the student realistically capable of mastering the demands of the situation right now? Does he have the skills necessary to meet the demands?

- If not, is it realistically possible to address the factors that are contributing to his difficulties right now so as to help him develop the skills that would make mastery of the demands of the situation possible? Could he successfully meet the demands if he was given some extra help?

Then the choices are the same: The unimportant ones should probably be avoided until the student can realistically handle more on his plate, the important but neither capable nor trainable ones can be altered (teachers already know tons of ways to do so for academic subjects—sometimes they have to give extra thought to how to do it for behavior), and extra training and support can be provided to the student for the important ones that appear within his reach.

In a user-friendlier school environment, adults read the warning signals and take quick action when these signals are present. As you've read, some inflexible-explosive students refuse to work when they're challenged by certain tasks. Danny would slump down in his seat, cross his arms, and announce that he could do no more (at those moments, he was probably right). Those were the warning signals. Danny was lucky: His teacher didn't push him when it was clear he was in vapor lock; instead, she tried to help him articulate what he was frustrated about. Sometimes Danny was able to explain the problem immediately; sometimes he couldn't. Sometimes his attempts to explain were spiced with inappropriate language (luckily, none of his classmates had virgin ears). Once in a while, Danny's agitation seemed to worsen when his teacher tried to help him. At these times, the teacher would back off, give Danny some time to chill, and come back to the issue when the odds of successful thinking improved. Over time, Danny began to trust that his teacher understood his tendency to get stuck and that she had a good sense about when he was apt to do so. He became better at verbalizing his difficulties and came to appreciate that the teacher seemed able to help him get unstuck. Often, before the teacher introduced an assignment to the entire class, she would quietly let Danny know she thought he'd have some trouble with it and reassured him that she would help him get started (this, of course, is an example of the sine qua non of proactive, front-end thinking: intervening *before* there's trouble). Sometimes she had another student help Danny.

Just remember, the burden for recognizing the early warning signs falls to the adults who are interacting with the student. It's critical to read these early warning signs and take action while there's still some coherence to work with to prevent these hints from turning into something much bigger, uglier, time consuming, and disruptive to the teaching process.

In a user-friendlier school environment, adults can interpret incoherent behaviors for what they really are: incoherent behaviors. The same theme as before: A lot of the words that come out of a student's mouth when he's frustrated are just mental debris. Your job is to get past the style of delivery and get to the actual message: HELP! Here's what the options may be:

Ring Out the Old

Lamar [*age thirteen*]: I'm not working on this assignment right now.

Teacher: Well, then your grade will reflect both your attitude and your lack of effort.

Lamar: I don't give a shit about my grades, man. I can't do this shit.

Teacher: Your mouth just got you a detention, young man. And I don't want students in my classroom who don't do their work. Anything else you'd like to say?

Lamar: Yeah. . . . This class sucks.

Teacher: Nor do I need to listen to this. You need to go to the assistant principal's office . . . now.

Bring in the New

Lamar: I'm not working on this assignment right now.

Teacher: Well, there must be something about the assignment that's hard for you. Let's see if we can figure it out.

Lamar: Forget it. . . . I can't do this! Just leave me alone! Damn!

Teacher: Lamar, listen a second . . . I know you have trouble with writing and spelling and you get very frustrated when you have to do assignments where you have to write and spell. Let's see if we can find a way for you to do the important part of the assignment—letting me know what you thought of the story you just heard, which is something you're very good at—without you getting all frustrated about the writing and spelling part.

Lamar: How?

Teacher: Well, maybe Darren would help you write down your thoughts. You could sort of dictate your thoughts to him.

Lamar: It's not worth it. No way.

Teacher: Oh, yeah . . . I sometimes forget you're embarrassed about how you spell. Why don't we have you and Darren do the assignment as a team and just make sure he does the writing part?

Lamar: You won't tell him I can't spell?

Teacher: Not if you don't want me to.

Lamar: Yo, Darren, let's do this thing together!

Better Still

Lamar: I'm not working on this assignment right now.

Teacher: Why, because it has spelling and writing? Are

you forgetting how we've decided to help you when you have to do assignments like that?

Lamar: Oh, yeah. Darren, get your butt over here and let's do this thing together!

In a user-friendlier school environment, adults can also understand how they themselves may be fueling a student's inflexibility-explosiveness. Are there any teachers out there who have, in one way or another, fueled a student's inflexibility and explosiveness? By misjudging his true capabilities? By assuming that compliance and flexibility and a tolerance of frustration are skills all children automatically possess? By believing that the student's difficulties all trace back to the home environment? By not being an especially wonderful role model for the other children in the class on how to interact with a child whose needs differ from the norm? By believing that a student's difficulties were totally accounted for by poor motivation, a bad attitude, or attention seeking? By creating a classroom environment that was distinctly *un*friendly? By communicating with the student in a manner that promoted an adversarial pattern of interactions? By expecting the administrative staff to pull out the sledgehammer every time a student was sent down to the office? OK, so maybe you hadn't read this book yet. Do it differently the next time.

In a user-friendlier school environment, adults try to use a more accurate common language to describe various elements of the student's inflexibility-explosiveness. Once again, the terminology we use for describing these students to themselves is crucial; since they don't necessarily have a well-developed vocabulary for describing themselves, they may start to use—and believe—our vocabulary. That can be good or bad, depending on the accuracy and usefulness of our vocabulary. I strongly encourage the use of certain terms to describe these children: *easily frustrated, difficulty thinking things through, difficulty shifting gears,* and so forth.

The vision thing. In my research, my colleagues and I have interviewed many teachers, so we know that different teachers have different visions of what they hope to achieve during a school year, both with individual students and with the class as a whole. Because teachers often spend more time with a child than his parents do, they can have a very potent impact. In this important respect, I envy teachers because they get a lot more time each day with the child, while I get him, on average, only for an hour a week. Can an inflexible-explosive student be a productive, respected member of the class? Is it possible for one teacher in a class of twenty to thirty students—many of whom also have special needs—to devote the time to meet the needs of such a student?

The concept of the responsive or prosocial classroom is an intuitively appealing vision that can help teachers answer these questions in the affirmative. This vision is described more fully in *Teaching Children to Care: Management in the Responsive Classroom,* by a teacher, Ruth Sidney Charney. Prosocial classrooms are those in which teachers place as heavy an emphasis on a "social curriculum" as they do on the academic curriculum (mathematics, reading, writing, and so on). In such classrooms, individual differences, such as great reading skills, poor writing skills, great athletic skills, poor spelling skills, and—why not?—inflexibility and low frustration tolerance—are celebrated and accentuated. Some of the same themes I discussed with regard to families apply to prosocial classrooms as well. Everyone in the classroom has strengths that can be used to help other students, and everyone has areas of weakness on which they need help. In the prosocial classroom, everyone gets what he or she needs. Thus, no one is singled out and stigmatized. Indeed, because of the emphasis on individual differences, students learn to help one another and learn that each classmate is an integral thread in the fabric of a community of learners. Planned and unplanned discussions (sometimes called "social conferences") focusing on social interactions are common. Teachers interrupt other ongoing lessons to process important social issues that have suddenly erupted in the classroom. Conflict resolution—compromising, negotiating, dealing with frustration, and generating alternative solutions to problems—can be an integral focus of such discussions. All students benefit from such discussions, but especially those who need it most. In such classrooms, teachers act as role models and provide frequent opportunities for practicing social interactions and helping one another, including peer tutoring, judicious seating arrangements, and cooperative learning.

I was in a prosocial classroom observing one of the inflexible-explosive children with whom I was working and was pleased to see him helping another student with her math (the child with whom I was working was strong in math). Several minutes later, it was time to switch activities, and, predictably, this boy became stuck. The girl he'd been assisting calmed him down, talked to him, and helped him move on to the new activity, all under the watchful gaze of the teacher. Then the girl, apparently recognizing that the stranger observing the class (me) might not have comprehended what was going on, came over to me and said softly, "He just gets a little frustrated sometimes." I nearly fell over.

This notion that all students have something important and valuable to contribute to a classroom is perhaps best described by my good friend and colleague, Dr. Robert Brooks, in his book *The Self-Esteem Teacher*. Bob is famous for asking, "What good are a student's strengths if no one knows about them?" Because of the severity of their behavior, inflexible-explosive students tend to be "defined" by this behavior, rather than by their many positive characteristics. A classroom is, without question, one of the ideal places to accentuate these positives—in other words, to focus on what Bob calls "islands of competence"—and build alliances between these students and adults. Alienation is the end result of the failure to do so.

So when teachers tell me that they'd like to be helpful to the inflexible-explosive student in their classroom but simply can't devote sufficient time to him, given the demands of their other students, I tell them about the prosocial classroom and how it's possible to get help from the least likely but most available and willing candidates: the other children.

Teacher: "I can't have different sets of rules for different kids. If I let one child get out of or get away with something my other students will want to as well."

First off, you probably have different expectations for different children already, so, as in families, in classrooms fair does not mean equal anyway. For example, while it would be nice if all your students could read at grade level, not every student can. So you probably make accommodations and provide assistance to those who can't. If one of the other students inquires about why a student is receiving such accommodations and assistance, you have the perfect opportunity to do some educating: "Everyone in our classroom gets what he or she needs. If someone needs help with something, we all try to help him or her. And everyone in our class needs help with something."

Talking the talk and walking the walk are no different when the developmental deficit is inflexibility and a low tolerance for frustration. So our response to the student who asks why an inflexible-explosive classmate is receiving some sort of special accommodations and assistance would be similar: "Everyone in our classroom gets what he or she needs. If someone needs help with something, we all try to help him or her. Because you're very good at handling frustration, I bet you could be very helpful to Johnny the next time he gets frustrated."

Children are actually pretty good at understanding the "fair does not mean equal" concept and at making exceptions for children who need help; in my experience, it's much more common that adults are the ones struggling with the principle.

Children are actually pretty good at understanding the "fair does not mean equal" concept and at making exceptions for children who need help; in my experience, it's much more common that adults are the ones struggling with the principle.

I've yet to come across a situation in which a child who typically behaved adaptively decided to behave inappropriately because accommodations were being made for an inflexible-explosive child in the classroom. As I already mentioned, it follows that punishing a child to set an example for, or to be fair to, the others—especially when there's no expectation that the punishment will be an effective intervention for the child being punished—makes little sense. In a community of learners, the academic or behavioral idiosyncrasies of one student are an opportunity for his or her classmates to help and learn, not to follow suit.

Except for some extreme circumstances—discussed in Chapter 13—inflexible-explosive children *can* function in regular education classrooms, often without a great deal of support. As you know, this usually requires a highly collaborative effort involving the child's parents, special education coordinator, school and outside psychologists, guidance counselor, principal, occupational therapist, speech therapist, and psychopharmacologist. All these adults should have important and unique pieces of information to add to the understanding of the child and treatment picture. The teachers know what they see at school and what they're doing to help the child be more successful and manage frustration more adaptively. The parents know best how things are going at home, can provide insight into the child's perceptions of how things are going at school, and often have a clear sense of which homework assignments are causing the greatest frustration. The special education coordinator knows what services the school system may be able to provide over and above those already being offered. The psychologist, speech and language therapist, occupational therapist, and guidance counselor can offer additional insight into specific areas of concern and the factors that seem to be fueling the child's frustration, whether through psychoeducational testing, neuropsychological testing, psychological testing, additional evaluations, or conversations with the child. The principal can open a lot of doors in a school; if special arrangements need to be made for a child, the principal is the person who can oversee the logistics. And you'll learn more about what a psychopharmacologist can do in the next chapter. It's important that the psychopharmacologist be an integral part of the process; I've never been enthusiastic about secondhand information, especially with regard to the adjustments required for optimal medical care. Why so few school systems have an established relationship with a consulting psychopharmacologist has always been a mystery to me.

With all that talent in the room, an understanding of a child and a consensus regarding a cohesive treatment plan—what the child is capable of right now, what issues are essential to address right now, what situations to avoid, what situations to adjust, what skills to build, when to increase supervision, and what treatments to bring to bear—should emerge, guided, I hope, by many of the principles in this book. Even if the initial treatment plan goes well, the whole crew should reconvene every two to four weeks or so to review the issues, adjust the intervention strategies, and revise the goals. Over time, if things go really well, fewer meetings will be necessary. Before the beginning of each school year, the new teachers should be brought into the loop to ensure the smoothest possible transition.

The tragic shootings that have occurred recently in our public schools underscore several important points: 1) there are many students who are capable of committing acts of unimaginable violence; 2) identifying children at risk for such acts must become a high priority and must involve the entire student body (most of the students who were involved in these shootings gave advance warning of their intentions to their fellow students); 3) we must redouble our efforts to find ways of ensuring that all students feel they are an important part of the social fabric of their schools; 4) if they are to remain in public school, such students will require the type of collaborative teamwork described above; and 5) suspending or expelling such students, or handing them over to the police, does not satisfactorily address their difficulties. Are any of the students who have committed these horrific acts inflexible-explosive? Not knowing any of them personally, I can't say for sure. Is it fairly clear that they weren't able to think of better ways to handle whatever problems they were having at school? The evidence is before us.

It is indeed the case that some inflexible-explosive children aren't able—despite everyone's best efforts—to stay safe within the school environment. More on this in Chapter 13.

Drama in Real Life
Edgar's Baskets

Edgar was an inflexible-explosive fourth grader with ADHD, expressive language delays, and sensory integration issues. Though he was apparently well medicated, he still struggled with tasks requiring sustained attention, particularly those involving written expression. His school system had provided him with a part-time classroom aide, with whom Edgar had worked reasonably well since second grade. But in the fourth grade, he was assigned to a different aide. Early on, he began refusing the help she offered, insisting that he didn't want to be different from the other children. As the school year progressed, Edgar's resistance to being helped worsened, and his academic performance and behavior suffered as a result. When his teacher and aide permitted him to try working independently on lessons with which Edgar had difficulty, he completed little work and tended to wander around the room, interrupting the work of other students. When his teachers insisted that he accept help, he often exploded and ended up in the principal's office.

Needless to say, Edgar's teachers, principal, and school psychologist were eager to see Edgar receive the help he needed. They were curious about the "basket thinking" that was helping Edgar explode a lot less at home.

"Basket thinking is a way of prioritizing our goals for children who aren't able to meet all our priorities," I explained. "It also helps us train some very important skills, such as compromising, perspective taking, and problem solving."

"How would we use this system with Edgar?" asked his teacher.

"Well, we need to think about what Edgar is capable of and what our priorities are for him," I said. "Because he's so easily frustrated, and has difficulties in so many areas—social, academic, and behavioral—we need to be really judicious about what we're asking of him. In Basket A are very high priorities that Edgar must do. In Basket B are important things on which there's some give and take. And Basket C includes things we're not going to pay much heed to right now. There are a lot of things that Edgar needs help with, but what are our top priorities?"

"He's so sensitive about looking different," said the school psychologist. "But I don't know if there's any way around it."

"He really needs the help," said the teacher. "Without it, he gets practically nothing done. I'm really concerned about whether he's learning anything."

"He won't even try to write in cursive," said the aide. "Even when he's willing to write, it's in print. And he refuses to show his work in math. He gets correct answers, but he can't tell you how he got them."

"I can't let him keep disrupting the classroom," said the principal. "I have other students in there whose learning is suffering because of Edgar's outbursts."

"I'm very worried about how his peers react to him," said Edgar's father. "He hasn't had a kid over to the house to play in a very long time."

"Well," I said, taking a deep breath. "That's quite an agenda. Edgar's not going to be able to work diligently on all these things at once. My opinion is that his accepting help is probably our top priority. He's not learning anything or getting any work done without help, and he's being disruptive to the learning of others when he's not working. I think accepting help goes in Basket A."

"I agree," said the teacher. "That would help us kill several birds with the same stone."

"Now, the last time I talked to Edgar," I said, "he told me that looking different isn't the only reason he doesn't like getting help. He said the help is provided at unpredictable times. So I think Edgar would experience the help as user-friendlier if it was provided on a predictable schedule."

"That's easy enough," said the principal. "We just have to decide what work he needs the most help on and what work he can do on his own. Then we can schedule the help during the times he needs it the most."

"Great," I said. "What work won't he even attempt without help?"

"Anything involving writing," said the teacher.

"Do we want the help provided in the class or outside the class?" I asked.

"He hates leaving the classroom," said the aide. "But he

hates having me work with him in front of the other kids even worse. And he gets less done when he gets help in the class."

"It sounds like leaving the room for help is in Basket A," I said. We then discussed a daily schedule for receiving help. We also agreed that writing in cursive was, for the time being, in Basket C. In other words, we concurred that it was more important to have Edgar complete more work than to have him write in cursive.

"How are we going to get him to accept the help without getting upset?" asked the father.

"Ah, that's where you come in," I said.

"Me? What can I do?" asked the father.

"If Edgar doesn't cooperate with leaving the room to get help as scheduled, these good folks at school are going to call you so you can talk to him," I said.

"Better you than me," the mother said under her breath.

"What would I say to him?" asked the father.

"You'll remind him of the necessity of accepting help and tell him you expect him to go work with the aide the minute he gets off the phone with you," I said. "If he balks, you'll let him know that he has no choice and that he won't be a happy camper if you get home from work and hear that he didn't accept the help."

"That's powerful," said the principal. "That's a lot more potent than my having him sit in the office."

"I can do that," said the father.

"I'd be surprised if he balks at accepting help from now on," I said. "But that's probably the only thing that's in Basket A for now."

"I don't know," said the teacher. "I think his saying mean things to other kids belongs in Basket A, too."

"I'm not sure about that," said the principal. "He's a very impulsive kid, and I'm not sure he's aware that he's saying something hurtful before he does it."

"What's he say?" I asked.

"Often he'll say, right out loud, in front of the entire class, that someone's answer was stupid," said the teacher. "Stuff like that."

"To tell you the truth, knowing Edgar, I'd probably put that in Basket B," I said. "My bet is that Edgar is simply being too 'honest.' He's certainly not very good at being politically correct."

The parents laughed. "There isn't a politically correct bone in his body," said the mother.

"My bet is that Edgar would be better off if we gave him different words for saying the same thing," I said.

"Like what?" asked the aide.

"Well, a lot of us would love to tell people that something they just said was pretty stupid," I said. "But what we say instead is something like, 'I'd like to add something to that answer.' I just don't think Edgar has the linguistic wherewithal to say what he's thinking in a way that isn't so honest. But I also think that's a lesson the entire class could benefit from—you know, not saying exactly what you're thinking but getting the point across in a kinder manner."

"So you mean we could have class discussions about the issue?" said the teacher. "I like that idea. But what should I do if Edgar says something mean in the middle of a lesson?"

"Stop the lesson briefly and suggest what you think Edgar—or anyone else—might have been trying to say," I said.

"This I can do," said the teacher. "I like it much better than sending him out into the hall."

"Sending him out in the hall would be fine if you thought he had good political skills that you were trying to motivate him to use," I said. "But, you're exactly right, sending him out into the hall doesn't help much if the real problem is that he doesn't have the words."

"OK, one last issue," said the aide. "What do we do about his wandering around the room?"

"Well, we're already doing something about that by having him do his most difficult work with you," I said. "But my sense is that Edgar needs more frequent breaks than other kids. I think we ought to give him a certain number of discretionary breaks every day, so he'll know we're aware of the fact that he needs breaks. And I think we need to identify certain things that are OK for him to do during these breaks. But he needs to know that the instant he disturbs other students, the break is over. Are there things in the room he can do during these breaks?"

"He loves the science center," said the teacher. "If he'd go to the science area and not disrupt other kids, I'd be very happy."

"Would the other kids mind that he gets breaks and they don't?" I asked.

"They're accustomed to his being out of his seat," said the teacher. "I think they'd be relieved just to have him not bothering them."

"I think we've got a plan," I said. "There are some

things we're not attending to right now. But I think we've identified the most important issues. By being very judicious about what's most important, and by thinking about how we're going to deal with each on the front end, I think the stage is set for things to go well."

"When should we meet again?" asked the school psychologist.

"Let's give this two weeks," said the principal. "We still need to present all this to Edgar. But I like the idea of having a plan to deal with these issues before he falls apart, rather than responding after he falls apart. Let's give it a shot."

Drama in Real Life
"Running on Empty"

"Dr. Greene, we can't let Casey keep running out of the room," the school principal said gravely. "It's dangerous, and we're responsible for his safety."

The principal was presiding over a meeting in March of Casey's first-grade year that included Casey's teacher, occupational therapist, guidance counselor, special education coordinator, parents, and me. Casey was blowing up a lot less often at home, but there were still some kinks to work out at school.

"Well," I said, "as you know, in some ways Casey's leaving the classroom is more adaptive than some of the other things he could be doing in response to frustration—like tearing the room apart. But, I agree, it's very important that he stay safe."

"What's making Casey act this way?" asked the classroom teacher. "What's his diagnosis?"

"Let's just say that Casey is very poorly self-regulated—like a lot of students with ADHD—and that he is easily frustrated and easily overwhelmed," I said.

"So why does he run out of the room?" asked the teacher.

"His running out of the room is probably a signal that his frustration with a particular task has become overwhelming," I said. "Does that correspond to your experiences with him?"

"Oh, absolutely," said the teacher. "He does fine on the things he's good at. But when he's confronted with something that challenges him or requires extra effort, he seems to become frustrated very quickly."

"I think we need to think about how to help Casey 'hang in there' when he's frustrated so he doesn't run out of the room," I said. "I also think we need to hone in on the specific situations that are causing him significant frustration, so we can decide which ones to avoid, which ones to adapt, and which ones to work harder on. But he may not stop running out of the room completely yet, so we may need a place where he can go to settle down when he does run out, so he doesn't end up in the parking lot."

The special education coordinator chimed in. "I think we should have consequences for leaving the classroom," she said. "I don't think it's good for the other kids to see him leave when he gets frustrated."

"Why, have any of the other kids expressed a desire or shown an inclination to leave the classroom when they're frustrated?" I asked.

"No," said the teacher.

"Do we think Casey is leaving the classroom because he'd rather be out in the hallway all by himself?" I asked.

"I don't think so," said the teacher. "He's always very eager to come back in as soon as he's settled down."

"Do we think that punishing him after he leaves the classroom will have any effect on his behavior the next time he's frustrated and feels the need to leave the classroom?" I asked.

"I don't know," said the teacher. "It's almost like he's in a completely different zone when he's frustrated."

"Then I'm not certain why we'd punish Casey for leaving the classroom," I said. "Especially if the main reason we're doing it is to set an example for the other kids."

"So what do you suggest we do when he gets frustrated?" glared the special education coordinator.

"I think most of our energy should be focused on what to do before Casey gets frustrated, not after," I said. "When Casey's frustration with a particular task or situation is predictable, we can anticipate his frustration and steer him around it. We can do things to make those tasks more doable for him, making them more stimulating, limiting the amount of time he has to spend on them, seeing if he'll accept help from one of his classmates. . . . I think there are a lot of ways to make these tasks more manageable for him. When the frustration is unpredictable, I think we need a place for Casey to go to calm down if your initial efforts to calm him down don't do the trick. I don't think he's at the point yet where he's able to talk things through when he's frustrated, although we're working on it. Luckily, he's pretty good at calming down on his own if we leave him alone for a while. We have to find ways to let him do so while still making sure he's safe. So for now, our top priority is to keep meltdowns to a minimum, even at the expense of his learning. It's the meltdowns that are getting in the way of Casey's learning anyway."

Things went reasonably well for Casey for the last few months of that school year. At the beginning of the next school year, the group reassembled, including his old and new teachers, reviewed what worked and what didn't the previous school year, and agreed to try to do more of the same, while focusing on helping Casey complete more schoolwork. Although we expected some rough moments as Casey adjusted to his new teachers and classmates, it wasn't until two months into the school year that he had his first series of meltdowns. The special education coordinator hastily called a meeting.

"We think Casey has regressed," the principal said. "He looks as bad as he did last school year."

"Actually, we think he looks a lot better than he did last school year," said Casey's father. "In fact, we were surprised he started off as well as he did. He was really looking forward to going back to school."

"I think we need to revisit the idea of consequences," said the special education coordinator. "Do you folks say anything to him about this behavior at home?" she asked the parents.

"Of course we do!" said the mother, a little offended. "We let him know very clearly that it is unacceptable, and he gets very upset because he knows that already. Believe me, this is being addressed at home."

"Is he melting down a lot at home?" asked the principal.

"We haven't had a major meltdown in months," said the

father. "We'd almost forgotten how bad things used to be."

"I still think Casey needs to know that at school, life doesn't just go on like nothing happened after he has a meltdown," said the principal.

"I agree," said the special education coordinator.

"What did you have in mind?" I asked.

"I think after he blows up, he needs to sit in my office and talk it over," said the principal. "And until he does, he shouldn't be permitted to rejoin his classmates."

"I don't think he's ready for that yet," I said.

"Well," said the special education coordinator, "whether he's ready or not, it's important that the other students see that we disapprove of Casey's behavior."

"His classmates don't already know you disapprove of his behavior?" I asked.

"We think we need to send a stronger message," the special education coordinator said. "We think he can control this behavior."

"I think he needs something more than a stronger message," I said. "Just remember, we shouldn't use consequences just for the sake of using consequences or just because it's the first intervention that comes to mind but, rather, because we believe consequences will put something in the forefront of Casey's brain that will help him refrain from doing things we wish he wouldn't do when he becomes frustrated. If we don't believe that will happen, consequences are likely only to make him more frustrated."

"We have to do what we think is right in our school," said the principal, ending the discussion.

Casey had a minor meltdown two weeks later. He was

escorted to the principal's office. The principal tried hard to get Casey to talk about his frustration. Casey couldn't. The principal insisted, setting the stage for a massive, one-hour meltdown that included spitting, swearing, and destroying property in the principal's office. Another meeting was hastily called.

"I've never been treated that way by a student!" said the principal. "Casey's going to have to understand that we can't accept that kind of behavior."

"Casey already knows that behavior is unacceptable!" said the mother. "Sometimes he can talk about what's frustrating him right away—and that's a recent development—but most of the time he can't talk about it until much later, so we have to give him some time to collect himself before we try to talk with him."

"I tried that," said the principal. "When he was in my office I told him that I wasn't going to talk to him until he was good and ready."

"How did he respond to that?" I asked.

"That's when he spit on me," said the principal.

"I guess that tells you that something about what you said made him more frustrated, not less," I said.

"You don't think having him sit in my office will eventually help?" asked the principal. "I'm very uncomfortable having him blow up and then watching him go happily out to recess and rejoin the other kids without there being some kind of consequence. I'm struggling with this."

"I think sitting in your office would work great if Casey experienced it as a place where he could calm down, rather than as a place where he's asked to do something he can't do yet— namely, talk about things immediately—or where he feels he's being punished for something he already knows he shouldn't have done."

"So why doesn't he just tell me he knows his behavior is unacceptable?" asked the principal.

"I don't think Casey can figure out why he behaves in a way he knows is unacceptable," interjected the father. "After this recent episode, he was very upset. That night, he practically begged me to give him medicine so he wouldn't act that way anymore."

The assembled adults were silent for a brief moment.

"But I can't give the other children in his class the idea that they can do what he does and get away with it," said the principal.

"I honestly don't think," I said, "that the students who are flexible and handle frustration well are going to starting exploding just because they see Casey getting away with it. They do need some help knowing the best ways to respond to him. And they're not going to see Casey getting away with anything—they're going to see that you take safety seriously and that you're doing everything you can to help Casey handle frustration more adaptively. I think Casey would be very willing to discuss how he can make amends for what he's done, so long as the discussion takes place later—when coherence has been restored—and so long as he perceives that you understand how hard this is for him."

Did Casey run out of the classroom again during the school year? Yes—to a designated desk in the hallway he knew was his "chill-out" area. Did he begin returning to the classroom much more rapidly after he left? Absolutely. Did he hit his principal again? No. Did he hit his classmates a few times? Yes—which placed him somewhere in the average for boys in his class. Did he continue to have trouble shifting gears? Of course. But his teacher demonstrated to Casey that she could help him when he became frustrated, and Casey thrived in her class. One day, I asked the teacher, "Do you think Casey's difficulties affect his relationships with his peers?" She replied, "Oh, I think he's well liked despite his difficulties. I think the other kids can tell when Casey's having a rough day, and they try to help him make it through."

12
Brain Chemistry

It's not possible to write a book about inflexible-explosive children without discussing medicine, if only briefly. My purpose in writing this chapter is not to provide you with all the ins and outs of psychopharmacological treatment of inflexible-explosive children and adolescents. Such a goal is well beyond the scope of this book and my own expertise. My goal is simply to orient you to some of the medications that may be used, either alone or in combination, to address the various pathways that may contribute to inflexibility and poor response to frustration.

Medications do have the potential to be enormously beneficial, as a means of both clearing the smoke and directly addressing the factors that are fueling a child's inflexibility and explosiveness. Indeed, there are some children who will not benefit substantially from the nonmedical approach described in this book until they've been satisfactorily medicated. Nonetheless, deciding whether to medicate one's child is often difficult; you'll need a lot of information—much more than is provided in this chapter. You may wish to read a new book by Dr. Tim Wilens, my colleague at Mass General, called *Straight Talk About Psychiatric Drugs for Kids*.

Ultimately, what you'll need most is an outstanding pediatric psychopharmacologist. You'll want one who:

- Is current on the latest research findings.

- Has a good working knowledge of the potential side effects of medications and their management.

- Is an expert on "combined pharmacotherapy"—the art of utilizing several medications.

- Is familiar with and recognizes the potential benefits of nonmedical interventions.

- Takes the time to get to know and understand you and your child up front and to listen to and address your concerns.

- Makes sure that you—and your child, depending on what's developmentally appropriate—understand each medication and its anticipated benefits and potential side effects and interactions with other medications.

- Tries to prescribe medications that have the fewest potential side effects and at the lowest possible dose and slowly moves on to higher doses or other medications with a greater potential for side effects only if the initial medication treatment plan fails to achieve its objectives.

- Monitors your child's progress carefully and continuously over time.

When I hear about children who have had extreme difficulties with medication, it is often because one of the foregoing elements was missing from their treatment.

All medications—aspirin included—have side effects. Fortunately, extremely serious side effects tend to be rare. Of course, regardless of their infrequency, side effects are always worrisome when they happen to your child. Your psychopharmacologist should help you weigh the anticipated benefits of a medication with the potential risks so you can make educated decisions. Although it's important to have faith in the psychopharmacologist's expertise, it's equally important that you feel comfortable with the treatment plan he or she proposes, or at least that you're comfortable with the balance between the benefits and risks. If you are not comfortable or confident in the information you've been given, you need more information. If your psychopharmacologist doesn't have the time to provide you with more information, you need a new psychopharmacologist. Medical treatment is not something to fear, but it needs to be implemented and monitored competently and compassionately.

Once an understanding of your child's specific difficulties has been achieved, medications may be considered as a means of addressing one or more of the various issues that may be compromising your child's capacity for flexibility and a tolerance of frustration, including inattention and cognitive inefficiency; hyperactivity-impulsivity; an irritable, agitated, or dysphoric mood; extreme mood instability; and anxiety or obsessive-compulsiveness. As you'll read, some medications address several of these issues at once. Not all the medications described here have been officially approved for use with children by the U.S. Food and Drug Administration, nor have many been studied extensively in use with children and adolescents, especially with regard to their long-term side effects.

Inattention and Cognitive Inefficiency

As you know, inattention and cognitive inefficiency are often seen in children with ADHD and can contribute to inflexibility-explosiveness both by fueling a child's cumulative level of frustration and by hindering a child's capacity to think things through and organize his thoughts. The mainstays of medical treatment for inattention and cognitive inefficiency are the stimulant medications. These agents are thought to increase the levels of the neurotransmitters dopamine and norepinephrine in the brain, either by blocking their re-uptake or stimulating their release. The stimulants, which have been in use for over sixty years, include well-known, well-studied medicines, such as methylphenidate (Ritalin) and dextroamphetamine sulfate (Dexedrine), and other agents, such as magnesium pemoline (Cylert) and a mixture of amphetamine salts called Adderall. Stimulants come in short- and long-acting preparations and are highly effective in improving attention and cognitive efficiency in many children.

In most cases, the side effects associated with stimulants tend to be mild, but they are worth mentioning. Two of the more common side effects are insomnia (especially if a full dose is administered after the mid- to late-afternoon hours) and loss of appetite, which can, over the long term, result in weight loss. In some children, stimulants may unmask or exacerbate existing tics (this circumstance may require adding a second medication to reduce the tics or reducing or discontinuing the stimulant medication). Stimulants may increase anxiety and irritability in some children—an often undesirable circumstance for inflexible-explosive children. The behavior of some children worsens when the stimulant medication wears off (a phenomenon called "rebound"), and this side effect is sometimes addressed by administering a half dose late in the afternoon to ease the child off the medication. Finally, particularly in adolescents, parents need to be aware of the potential for abusing stimulants.

Hyperactivity and Poor Impulse Control

Hyperactivity and poor impulse control can also compromise a child's capacity for flexibility and frustration tolerance. Some children are so hyperactive that they have difficulty adapting to even minor frustrations. And an impulsive cognitive style may hinder a child's capacity to think through options for dealing with a specific frustration. Several classes of medications are used to improve self-regulation in children, and stimulants are, again, often the agents of first choice. However, in some children, untoward side effects, the lack of a positive response to the stimulants, or complicating conditions may require a consideration of alternative medications for enhancing impulse control and reducing hyperactivity. The tricyclic antidepressants, which include agents like nortriptyline (Pamelor), desipramine (Norpramin), imipramine (Tofranil), and clomipramine (Anafranil), may be quite useful. An advantage of tricyclics is that they provide twenty-four-hour coverage and typically do not interfere with sleep. The tricyclics, especially nortriptyline and desipramine, have also been found to have positive effects on attention span and are the best studied of the antidepressant medications used in the treatment of childhood ADHD.

The side effects of these medications are also worthy of mention. One of the rare, but more serious, side effects of the tricyclics is cardiac toxicity, which often requires that children medicated with tricyclics undergo initial and then periodic electrocardiograms. There are a variety of additional potential side effects that may not be well tolerated by children, including dry mouth, weight gain, sedation, lightheadedness, and constipation.

An atypical antidepressant called bupropion (Wellbutrin) has also been used to ameliorate hyperactivity-impulsivity in children with ADHD. The side effects of bupropion tend to be better tolerated than those of the tricyclics. Bupropion may also improve mood and reduce irritability and anxiety. Thus, it may be especially useful in children whose ADHD-related concerns are complicated by mood and anxiety issues. Bupropion may increase the risk of seizures; exacerbate tics; cause insomnia, nausea, headache, constipation, tremor, and dry mouth; and initially increase agitation.

The antihypertensives, including clonidine (Catapres) and guanfacine (Tenex) are also commonly used to reduce hyperactivity and impulsivity, but may be less effective with inattention. Antihypertensives can also be effective in reducing tics and aggression. In addition, because of their sedating effect, antihypertensives have been used in children who tend to be overaroused at bedtime. However, in some children, this sedation may be a problem during daytime hours and is sometimes manifested in the form of heightened irritability. In inflexible-explosive children whose difficulties seem fueled by mood issues, this increased irritability is undesirable. Side effects can include headache, dizziness, nausea, constipation, and dry mouth.

Irritable, Agitated, Dysphoric Mood

As you know, mood issues can contribute to inflexibility-explosiveness. Children whose mood is chronically irritable, agitated, or dysphoric often respond to even minor frustrations as if they were major obstacles, in the same way many of us do during transient periods of irritability and agitation. Mood issues in children are often treated with a group of medications called selective serotonin re-uptake inhibitors (SSRI antidepressants), which include agents such as fluoxetine (Prozac), sertaline (Zoloft), paroxetine (Paxil), and fluvoxamine (Luvox). These medications are newer to the market than the tricyclic medications discussed earlier, and their popularity is owed to their effectiveness in easing a wide range of symptoms, including dysphoric mood and irritability. The SSRIs are not thought to be useful for treating the core symptoms of ADHD; indeed, in some children, they can actually cause behavioral activation, sometimes to the point of inducing manic behaviors. While the potential side effects associated with the SSRIs are thought to be better tolerated than those associated with tricyclics, they are noteworthy. The most troublesome untoward side effects include nausea, weight loss or weight gain, anxiety, nervousness, insomnia, and sweating.

Extreme Mood Instability

The term *mood instability* is often used to describe extreme levels of irritability and agitation and frequent aggressive outbursts. You may recall that some clinicians prefer the terms *bipolar disorder* or *manic-depression* to describe such children; in other circles, "really bad ADHD" or "ADHD with depression" are the labels of choice. Among researchers, these labels are still an issue of some controversy because of the overlap between the symptoms of ADHD and bipolar disorder. Regardless of the label one chooses, in my experience—and the research of my colleagues at Mass General confirms this impression—children with ADHD who also meet the criteria for bipolar disorder exhibit more severe behavior, are more impaired psychosocially, and have poorer long-term outcomes than do those with ADHD alone.

Although several of the medications described here may be used with such children, a class of medications called mood stabilizers—which broadly includes medications like lithium carbonate and anticonvulsants, such as carbamazepine (Tegretol), gabapentin (Nerontin), and valproic acid (Depakote)—can be quite effective. The mood stabilizers may be less effective in children who are predominantly dysphoric. Indeed, because these agents may produce drowsiness or fatigue, they may actually increase irritability in some children. All four provide twenty-four-hour coverage and do not affect the sleep of most children. The potential side effects of these agents are worth noting. Lithium can cause sedation, nausea, diarrhea, thirst, increased urination, mild tremor, and weight gain and must be monitored closely and continuously. Valproic acid and carbamazepine may cause sedation, nausea, diarrhea, heartburn, tremor, and weight gain. Valproic acid can also cause liver toxicity, and carbamazepine can be associated with a decrease in the white blood cell count and aplastic anemia, so the use of these agents requires periodic blood work. Gabapentin is thought to have milder side effects and does not require periodic blood monitoring.

In some instances, antipsychotic medications—which include newer medications, such as risperidone (Risperdal) and olanzapine (Zyprexa)—may be used to control extremely aggressive and explosive behavior. These medications have prompted much enthusiasm because they tend to be better tolerated than more traditional antipsychotics. However, these agents are associated with sedation and weight gain, and may be associated with extrapyramidal symptoms, such as eye-rolling, rigidity in the limbs, fixed facial expression, blank emotions, and involuntary movements. These latter side effects can be quite serious and should be described by your psychopharmacologist.

Anxiety and Obsessive-Compulsiveness

Anxiety can also set the stage for inflexibility-explosiveness, especially when it interferes with a child's capacity to think things through or causes the child to experience feelings over which he has little control. Several classes of medication may be used to address various forms of anxiety in children, including the SSRI antidepressants, which I discussed earlier, the anxiolytics, which include medications such as buspirone (BuSpar), and the benzodiazepines, which include medications such as clonazepam (Klonipin). In addition to helping with anxiety in general, the SSRIs can also be useful in the treatment of obsessive-compulsive disorder (OCD). In my experience, the OCD label is sometimes applied to children who are strikingly inflexible but do not actually satisfy the criteria for the full OCD syndrome. The SSRIs can help such children be more malleable, and this, of course, is a desirable effect in inflexible-explosive children. Clomipramine (Anafranil), one of the tricyclic antidepressants you read about before, is also commonly used to treat true OCD.

❄

No one wants to see a child medicated unnecessarily, so a conservative approach to medication is appropriate. But it's also worth repeating that in some inflexible-explosive children, medication is an essential—sometimes indispensable—component of treatment. In many instances, medications are combined because the child's difficulties have not been satisfactorily improved by a single agent. In other instances, a medication may be added to counteract the undesirable side effects of another medication that is having otherwise beneficial effects.

At the risk of redundancy, it should be clear that the most crucial component in the psychopharmacology picture is a competent, clinically savvy, attentive, available psychopharmacologist. But a psychopharmacologist can't treat your child successfully unless he or she receives accurate information from you and your child's teachers about the effects of the prescribed medications. When all relevant adults work in concert with the psychopharmacologist, side effects are handled more efficiently and adjustments made more responsively.

A discreet approach to medication is also recommended. Most children aren't eager for their classmates to know that they're receiving medication for emotional or behavioral purposes. And there's a temptation for parents to keep school personnel in the dark about their child's medication status as well. True to the collaborative spirit required for intervening effectively with inflexible-explosive children and because the observations and feedback of teachers are often crucial to making appropriate adjustments in medications, I generally encourage parents to keep relevant school personnel in the loop on medications. If there's no way to keep a child's classmates in the dark, it's often necessary to educate the classmates about individual differences (asthma, allergies, diabetes, difficulty concentrating, low frustration tolerance, and the like) that may require medicinal treatment.

Will your child be taking medication the rest of his life? That's hard to predict. In general, the chemical benefits of these agents endure only as long as the medication is taken. Nonetheless, in some children, the behavioral improvements that are facilitated by medication persist even after the medications are discontinued, especially if a child has acquired new compensatory skills. Ultimately, the question of whether a child should remain on medication must be continuously and collaboratively reviewed.

Finally, as a practical matter, you may be wondering if a child's compliance with medication is in Basket A. Most of the younger children I work with don't object to taking medication unless they have trouble swallowing pills, and adjustments for this problem are easy to make (for example, some medications come in liquid form, and pills can be crushed and mixed with food). When a child objects strongly to taking medication—and, in my experience, this circumstance is far more common in preadolescents and adolescents—it's usually a sign that the child is objecting to the side effects (drowsiness, dizziness, and nausea seem to be at the top of the list), feels stigmatized or embarrassed about having to take medication at school, wants to become a more active participant in discussions about his own care, or has grown tired of being treated like a patient. Whatever the issue, Basket B—where things can be discussed, processed, and negotiated—is where the issue usually lands. In other words, forcing pills down the throat of an inflexible-explosive fourteen year old is not a successful long-term strategy, nor is hospitalizing a child who refuses to take his medication (although the latter action is necessary under extreme circumstances). In the vast majority of cases, children are willing to take medication if their concerns are taken into account and adjustments are made in accordance with their needs.

13
Change of Venue

Just in case you didn't notice, I'm optimistic about the capacities of inflexible-explosive children to begin responding to frustration more adaptively. Yes, it's often difficult to create user-friendlier environments at home and at school. Yes, sometimes old habits and knee-jerk, back-end interventions die slow deaths. And yes, sometimes it takes a while to get the psychopharmacology piece squared away. But these are resilient kids—they do respond to being understood and to good treatment.

Most of the time. There are, unfortunately, children who do not have access to, refuse to participate in, or do not respond as favorably to treatment and who continue to behave unsafely at home, at school, and or in the community. Many started a downward spiral early, became increasingly alienated, began exhibiting more serious forms of societally inappropriate behavior, and began to hang out with other children who have come down a similar path. Once alienation and deviance become a child's identity and a means of being a part of something, things are a lot harder to turn around. Many such children ultimately need treatment that is more intensive than the kind that can be provided on an outpatient basis or in a regular education setting. In this chapter, we'll look at what is done under such circumstances.

First, a warning. Society—schools, the justice system, social service agencies, the mental health profession—isn't well prepared to help these children. As you've read, although school systems establish the priorities of safety and respect for authority, they often don't have high-quality programs for students who may have difficulty adhering to expectations in these domains. And many of the alternative day-school placements run by school systems or state agencies still have motivational programs as their primary therapeutic ingredient. The police and courts aren't equipped to provide the type of monitoring and intervention needed by many families. Often, the best the judicial system can do is hold the threat of a significant back-end consequence over a child's head. Many social service agencies are overwhelmed; the problems of an inflexible-explosive child and his family may pale in comparison to the problems of other children and families that are referred to and followed by them. The mental health profession isn't especially effective in working with individuals who won't come in for treatment or whose needs require attention outside the boundaries of a fifty-minute session in a therapist's office. And managed care issues are sometimes a significant obstacle. Finally, most states don't devote tons of money to the development of quality services for children whose inflexible-explosive behavior prevents them from fitting into the mainstream. That's why there's a building boom in new jails.

Once alienation and deviance become a child's identity and a means of being a part of something, things are a lot harder to turn around.

After all else has been tried—therapy, medication, and even alternative day-school placements—what many of these children ultimately need is a change of environment. A new start.

A way to start working on a new identity. One way to give them this new start is by placing them in a residential facility. As horrible as that may sound, there are some outstanding residential facilities in the United States that do an exceptional job of working with such children. Most of the six or seven children I've ended up placing in residential settings came out a lot better than they went in, and most maintained their gains once they went back home.

The big problem: These facilities cost a lot of money—as much as $80,000 a year. For the vast majority of families, that's an obscene amount of money. No school system, department of mental health, or social service agency is eager to part with that kind of money, either. And the political winds in many states are making it even more difficult to place children in these facilities when necessary.

Let's put money aside for the moment to dispel a few myths. The better residential facilities have excellent academic programs, so a child's chance of being accepted to college isn't sacrificed. Although many of these facilities have a behavior management component for maintaining order, the better ones also have a strong therapeutic component through which many of the thinking and communicating skills described in the preceding chapters can be developed. Many also have a family therapy component (remember, the goal is for the child to return to his own home and community). As you may imagine, I'm generally not enthusiastic about residential programs whose primary agents of change are large human beings who make sure your child knows who's boss. I've worked with several residential facilities on developing programs that are consistent with the alternative approach described in this book.

The prospect of placing one's child in a residential facility can feel like a nightmare to many parents, although parents who have been living a nightmare at home are often more open to the idea. Our instincts are to keep our families together, even when they're being torn apart. Our instincts are to keep our children under our supervision, even when our supervision is no longer sufficient. We don't like to feel like we're throwing in the towel, even when all the evidence suggests that we cannot provide everything a child needs. We don't like asking someone else to take care of our child, even when we think it may be for the best. So our every instinct is to hang on, tough it out, and try something else. A new drug. A new therapist. A new program. A new school. A new book.

Back to money. If a child is acting out in school, it is sometimes possible to convince or compel a school system to pay for a placement outside the school system. Under the Individuals with Disabilities Act, a federal law (and it's a good one) that applies to public schools, school systems are obliged to place their students in the "least restrictive" setting appropriate to their needs; residential placements are considered the *most* restrictive setting, so such placements are generally held out as a last resort. School systems vary widely in their resources for children who need more than the mainstream can provide. If it becomes apparent, after adjustments and accommodations are made, that a regular education program is not sufficient for your child's needs, the first consideration in many school systems is a classroom aide. The next step is often a special classroom for children with behavioral problems. A day-school placement outside the school system is frequently the next step. If these alternatives fail to achieve the desired effect, a residential program may become a more serious consideration. In some cases, if it's sufficiently clear that these intermediate steps are insufficient for a child's needs, a residential program may be considered earlier in the process.

In general, public school systems require a student to pass the following litmus test before a residential facility is considered: (1) the child's current program is thought to be inappropriate for meeting his educational needs (or the child's emotional issues are preventing his educational needs from being met); (2) there is no program within the system appropriate for the child's needs; and (3) the child's needs require the intensity or safekeeping of a twenty-four-hour program. As you may imagine, this is a pretty subjective litmus test, and it's fairly common for parents and school systems to come to radically different opinions on these issues. Some parents end up hiring lawyers or educational consultants who are familiar with the special education laws to help with the persuading. I've not seen many parents successfully negotiate this process alone.

Your other option is to ask to have your child placed in the custody of the state. The state then decides what the best facility for your child may be. Often, this decision is based more on funding and the availability of space and less on what a child actually needs. Unfortunately, the parents who end up pursuing this route are often those who lack the financial resources or wherewithal to advocate for themselves. Raising an inflexible-explosive child is difficult enough without the added burdens of financial hardship, dangerous neighborhoods, poor role models, and so forth. There are some organizations that can be helpful to such parents, and these organizations are listed in the Additional Resources and Support section at the back of this book.

If you end up seriously considering placing your child in a residential facility, try to visit any of the programs to which you're thinking of sending him. Make sure you feel comfortable with the staff, the philosophy of the program, and the other children at the facility. Make sure the facility has lots of experience working with children whose profiles are similar to your child's. Make sure the staff of the facility are open to your ideas about your child.

Placement won't be forever. With luck, only a year or two. That gives you some time to get your house in order. To have your child in a controlled, safe environment where he can learn how to think more flexibly and handle frustration more adaptively. To get his medication straight. To help him come home. It's not the end of the world. It can be a new beginning.

Drama in Real Life
Hard Times

You met Jennifer—star of the "waffle episode"—in Chapter 1. By the end of fifth grade, her teachers were just beginning to see signs that all was not right with Jennifer. Jennifer began to miss more school, usually because of some mild physical ailment. Some days she simply refused to go. Sometimes her mother would try to force Jennifer to go to school, subjecting herself to a barrage of verbal and sometimes physical attacks. A truancy officer had come to the home on several occasions; during one visit, Jennifer locked herself in the bathroom and refused to come out. Her teachers began to comment on the forced, "wooden" quality to her social presentation and noted that she came off as very bossy. Several new combinations of medication had been tried, with little change.

Our therapy sessions often went something like this:

"Jennifer," I'd begin, "from what I hear, you got very frustrated about something at home yesterday. Maybe I can help out."

"WHAT DID YOU TELL HIM!" she'd scream at her mother.

"Well, he knows I called him from a pay phone while you were tearing the house apart," the mother replied.

"WHY DID YOU TELL HIM THAT?" Jennifer would thunder, kicking her mother's chair.

"Because he can't help us unless he knows what goes on in our house," the mother would reply with some irritation.

"I DON'T WANT HIS HELP!"

"I do," the mother would say, "because I think we can get along better than we do."

"THEN YOU COME HERE BY YOURSELF! IF YOU KEEP TALKING ABOUT ME, I'M LEAVING THE ROOM!"

"Jennifer," I'd interject, "I was just interested in what you were so frustrated about . . . "

"I WAS JUST INTERESTED IN HAVING YOU SHUT UP!"

". . . before you started tearing the house apart."

"I WASN'T FRUSTRATED! THE ONLY THING THAT'S FRUSTRATING IS THAT YOU AND MY MOTHER AND MY FATHER ARE ALL FUCKING ASSHOLES!"

"Let's talk about school, then," the mother said, trying what she thought would be a less inflammatory topic.

"SCHOOL IS NONE OF HIS FUCKING BUSINESS EITHER!"

Sometimes we'd get some glimmer of hope that Jennifer might be able to participate in a meaningful conversation about her difficulties or be capable of some kernel of insight, but not for long and not often.

"Do you think she's ever going to 'open up'?" her mother would ask me privately.

"I don't know," I'd say. "The only emotion we ever get from Jennifer is anger. It's hard to tell what other feelings she has or whether Jennifer has any meaningful insight about her feelings because she becomes so anxious and frustrated the instant she's uncomfortable talking about something."

"She's been like that since she was born," said the mother. "It's always been hard to tell what she feels. Do you think she even knows what she feels?"

"It's hard to tell," I said.

"Can you imagine?" asked the mother. "All I've ever wanted from this kid is some sense of what's going on inside her head . . . some sense that she actually cares about the people around her. . . . "

"If she was the least bit receptive to seeing a different therapist, we could see whether she might be more open with someone else," I pondered.

"She's been to almost a dozen therapists already!" the mother reminded me. "Believe it or not, this is the most 'open' she's ever been, if you can call it that."

"That may be true, but we're not getting very far," I said. "Given her propensity for physical aggression, we don't have tons of time to figure it out. If she can't at least be a lot safer than she's being right now, I'm afraid we're going to have some very difficult decisions to make. I have this sinking feeling that starting middle school next year is only going to magnify her level of discomfort and make things even tougher all the way around."

This feeling turned out to be prophetic. When Jennifer entered the sixth grade, things headed downhill fast. She had difficulty adapting to the heightened social pressures of middle school and rapidly alienated many of her peers (when she perceived that a classmate disliked her, she'd spread vicious lies about the child). By winter, the police had to be called to the home several times because of her violent behavior. Jennifer continued to have massive difficulties engaging in any discussion that emphasized the necessity of safety or was aimed at helping her develop some basic skills for dealing with frustration. Her mother called me from a pay phone on more than one occasion: Jennifer was destroying the house; Jennifer's siblings were in the car outside the phone booth, scared to death. Despite everyone's best efforts, she was deteriorating.

In one session, Jennifer's mother tried to talk about a problem Jennifer had with a classmate at school during the week.

"Jennifer, Mrs. Johnson called me from school," the mother began. "I think it would be a good idea to talk with Dr. Greene about what happened between you and Eileen this week."

"I CAN'T BELIEVE SHE CALLED YOU!" Jennifer screamed, turning red. "IT'S NONE OF YOUR FUCKING BUSINESS!"

"We can't help you if you won't let us," said Jennifer's father.

"I DON'T NEED YOUR FUCKING HELP!" Jennifer screamed, becoming more agitated. "IF YOU KEEP TALKING ABOUT THIS, I'M LEAVING."

"Jennifer," said the father, "in the last month your mother has called me from a pay phone while you tore up the house, the police had to come to our house because the neighbors could hear you screaming at your mother, and — "

"I TOLD YOU I'M NOT LISTENING TO THIS!" Jennifer screamed. With that, Jennifer picked up a paperweight from the end table and hurled it, barely missing her mother's head. The father leaped out of his seat and wrestled his daughter to the ground. I called for hospital security.

"You cannot hurt your mother!" the father yelled. Jennifer tried mightily to escape his grasp.

Security arrived, and Jennifer and her parents were escorted to the emergency room. Jennifer calmed down quickly; ultimately, the emergency room staff did not think it was necessary for her to be hospitalized. The parents and I met the next day.

"I don't think you guys are going to be surprised by what I'm about to say," I began. "We've been at this for quite a while now. Jennifer's been on just about every psychiatric medication known to mankind. I'm afraid it's becoming obvious that we are not going to achieve our goals through outpatient treatment. I think we should consider looking for a placement outside your home."

The mother began to cry. She seemed relieved. The father wasn't quite ready.

"Maybe if we tried a new therapist," he said. "Maybe you're not the one."

"That could be," I said. "Don't forget, you've been to many different therapists already. Jennifer is unsafe, and you guys and your other children continue to be traumatized."

"Maybe she needs a female therapist," said the father.

"She's had a ton of female therapists," said the mother, tearfully.

"What kind of place are we talking about?" asked the father.

"A place with kids just like Jennifer," I said. "Kids who can't seem to deal with frustration, who become violent, who need a safe place to work on things."

"You mean a place with a lot of crazy kids running around?" the father asked.

"No," I replied. "Far from it. Anyway, it would be a good idea for you to visit some of these places. I think you'd be pleasantly surprised."

"She'll leave if she's put in a place she can escape from," the mother said.

"There are places that are prepared for kids who are likely to try to split," I said.

"She won't go on her own," the father said.

"Some kids have to be brought to these facilities involuntarily, sometimes in an ambulance," I said.

"How long will she have to be there?" the father asked.

"If we're lucky, only a year or two," I said. "It depends on how she progresses."

"How much do these places cost?" the father asked.

"A lot," I said.

"Who pays for it?" the father asked.

"That's something we'll need to explore," I replied. "Her teachers are well aware that her emotional issues are preventing her from learning, and her guidance counselor has been hinting that she doesn't think Jennifer is able to function in a regular education setting right now. So it's possible her school system will see the need for, and be willing to fund, a twenty-four-hour placement."

The father ran out of questions. He sat quietly while his wife cried. Tears began welling up in his eyes.

"This is my daughter you're talking about here," he said.

"I know."

"We love her."

"Yes, you do."

"We . . . uhm . . . we can't . . . live like this anymore."

"I agree."

The parents tearfully hugged each other and said nothing for several minutes. Finally, the father wiped his eyes and said, "Tell us what we need to do."

14

Children Do Well If They Can

When our daughter was born, my wife and I felt strongly that she should be breast-fed. When things didn't go quite as well as planned, I was struck by the tremendous disparity of opinions and advice that began flying our way.

"You guys need to let the baby know there's a feeding schedule very early on. You don't want to spoil her."

"You should feed her whenever she wants. Maybe she's just a baby who likes to breast-feed every half hour."

"Feed her whenever she wants, and you'll never know a moment's peace."

"Babies know what they're doing. Just relax."

"Babies don't eat when their mothers are too nervous."

"What that baby needs is some fennel tea—her stomach is too upset for her to eat."

"The problem is that there's simply not enough milk."

"The problem is that the milk is too light."

"You need to make sure she stays awake while she's eating from the breast. Otherwise, she won't consume enough at each feeding and she won't have the energy to breast-feed the next time."

"You should let the baby cry for awhile when she's hungry. It's good for lung development."

What did we know? We'd never done this before. Fortunately, we had some decent feeding alternatives, such as formula, available to us. So if we had to forgo our breast-feeding vision, at least we'd know our child would be relatively well nourished.

Parents of inflexible-explosive children are usually similarly unprepared. Most parents have never had an inflexible-explosive child before. As you've read, such children are not what most parents hope for and dream of when they envision having a child. What a disparity between vision and reality! When faced with such a circumstance, some parents diligently pursue their original vision and apply the same parenting strategies that were effective for them or their other children. And who can blame them? There has been no revised vision for them to work toward and no alternative technology to apply in the pursuit of this different vision. I hope that this book has helped fill the void on both counts.

The vision and strategies described in the preceding pages are guided by a simple philosophy that has evolved over the years in my work with all different kinds of children: **Children do well if they can. If they can't, we need to figure out why, so we can help.** This philosophy is in the forefront of my brain when I'm listening to parents' first description of the problems that have prompted them to seek help for their child. It energizes me to move beyond the standard motivational explanations and to identify the deficits in skills or neurobiochemical issues that underlie the child's difficulties. It reminds me that if a child *could* interact with other people in a more adaptive manner, he *would*. The vast majority of children do not get their jollies by making themselves and those around them miserable.

As you now know, my sense is that poor motivation is perhaps the most overrated, overinvoked explanation for why children do not meet our expectations. Unfortunately, for many adults, it tends to be the knee-jerk explanation. My advice: *Hold on to your knees*. For the most part, children are already motivated to do well.

Indeed, poor motivation is only one possibility within an entire universe of explanations for why a child may be inflexible and explosive. The explanation I've been advocating in this book is that, because of a variety of factors, some children are compromised in their capacities to be flexible and tolerate frustration adaptively. Compliance is a skill that does not come naturally to all children.

And being a better motivator is only one possibility within an entire universe of things adults can do to help. There's a wide range of additional parenting skills that are even more important: understanding who your child is, what his capabilities are, and under what circumstances his relative inefficiencies are likely to cause him difficulty; being sensitive to the ways your child communicates that he's stuck; and being responsive to these difficulties in a way that your child experiences as helpful and that fosters a parent-child relationship that is rewarding to you both. One of the most important roles we play in the lives of children is to set our expectations just within reach continuously throughout their development. When children become capable of clearing higher hurdles, we can start raising our expectations, slowly but surely.

Your new vision for your child? That's up to you, but I'm hoping that he'll come to view you as an ally, rather than as the enemy. That he'll eventually be interested in understanding himself a little better. That he'll start handling frustration more adaptively, communicate more effectively, and participate more actively and independently in the process of problem solving. That he'll feel better about himself and his relationship with you. That he'll feel more optimistic and competent and start to show the world his wonderful qualities without being tripped up by his inflexibility and explosiveness.

Your new vision for yourself? That, too, is up to you, but I'm hoping you'll be able to get back in touch with why you became a parent in the first place. That you'll feel like you finally understand and can communicate with your child. That you'll be able to accept his limitations and feel confident that you can move toward a more productive, loving relationship. That you'll relinquish goals you weren't ever going to achieve and feel like you can finally be who you really are when you're with your child. That realistic optimism will feel good.

You must be curious about what happened to many of the children and parents you met in the preceding pages. Casey's parents continue to implement many of the strategies described in this book, and medicines that have been prescribed to help Casey think more clearly and explode less rapidly have worked well. Casey hasn't melted down at home in many months. He still has the occasional meltdown at school, but his teachers are much better at predicting when he is likely to become frustrated and knowing how to respond in a way he experiences as helpful.

Danny never hits anymore. His mother is much better at helping him find different words to express his frustration, and Danny has become much better at using these words. He still swears occasionally, but things are so much better that his mother barely notices. Danny's medication has been adjusted several times; at present, he's doing fairly well on an antidepressant and a mood stabilizer. I still meet with him and his mother every other week. Most weeks, they walk in smiling.

Eric didn't come for many more sessions. His parents had a lot of trouble getting to the sessions and were really hoping someone could help him in individual therapy. I'm more of a family therapy guy, and I didn't believe I could give Eric tools in individual therapy that would be potent enough to overcome the frustrations he faced in his different environments.

Anthony's anxiety over the weather has waned; we're working on giving him cognitive tools that he'll be able to apply to *any* anxiety. He has a lot of trouble hanging on to new ideas, so things are taking a little longer than usual. But he's progressing.

Mitchell and his parents caught on to their rules of communicating—no sarcasm, no one-upmanship, and no mind reading—faster than I thought they would. They're communicating much more effectively these days. They're actually able to discuss things with one another and resolve difficult issues. Mitchell began trusting me enough to agree to a trial of an antidepressant medication, to which he responded very well. We still need to get him up and running at school. All things in due course.

Jennifer is off medication completely and in a new residential facility. The first facility—where she stayed for over a year—had an elegant point system, and Jennifer did behave herself, but the individual and family therapy components of the program weren't as strong as we had anticipated. Although it's doubtful that she'll ever live at home again, I'm still pretty optimistic about her ultimate outcome. I suspect that Jennifer will have her share of troubles as an adult. But she'll make a great attorney if we can help her survive adolescence with her self-esteem intact.

Her parents have finally been getting some hints that Jennifer has some strong feelings besides hostility. At the intake interview at her new residential facility, the interviewer asked Jennifer to say what she liked most and least about each parent. She had no trouble articulating what she liked least. But when it was time to talk about what she liked most, she began to cry, got out of her chair, and hugged both parents.

Finally, I saw Helen, the girl who needed the rudimentary vocabulary for feelings, a few weeks ago for the first time in many months. Her parents began the session by saying, "We don't really know why we're here today—Helen hasn't had a major blowup in a very long time." So we spent the session reminiscing about some of Helen's old episodes.

"We've got a great story to tell you," chuckled the mother. "This past Halloween, Helen decided she wanted to dress up as Selena—you know, the singer. Well, I spent a chunk of money buying material and then all kinds of time sewing her this incredible costume. It's the afternoon of Halloween and Helen comes to me and says, 'I think you're going to be mad at me. . . . I want to wear the costume I wore last year.'"

"Wow," I said. "She really said, 'I think you're going to be mad at me'? Now there's some anticipation and some empathy and some language we weren't hearing a year ago. What did you do?"

"I let her wear her old costume," said the mother.

"Not a Basket A issue, eh?" I asked.

"No way," said the mother. "But she's been handling Basket A issues really well, too."

"Lately, when she gets confused about something," added the father, "she's been saying things like 'I don't know what you mean' or asking us to explain things more clearly."

"Wow!" I said. "She's really been able to think that clearly when she's frustrated?"

"You bet," said the father.

"You guys and Helen have really come a long way," I said. "I'm delighted for you."

Children do well if they can. If they can't . . . well, now you know the rest.

About the Author

Ross W. Greene, Ph.D., is Director of Cognitive-Behavioral Psychology at the Clinical and Research Program in Pediatric Psychopharmacology, Massachusetts General Hospital, Boston, where he specializes in the treatment of inflexible-explosive children and adolescents and their families. He is also Assistant Professor of Psychology in Psychiatry at Harvard Medical School and Consultant in Psychology at McLean Hospital in Belmont, Massachusetts.

Dr. Greene's current research focuses on the classification, longitudinal study, and treatment of social impairment in children with ADHD; the treatment of inflexible-explosive children; and the influence of teachers' characteristics on school outcomes for elementary school students with ADHD. He has authored and coauthored numerous articles, chapters, and scientific papers on behavioral assessment and social functioning in children, school- and home-based interventions for children with ADHD, and student-teacher compatibility. He is on the editorial boards of the *Journal of Clinical Child Psychology*, *Cognitive and Behavioral Practice*, and *Journal of Psychoeducational Assessment*.

Dr. Greene received his doctorate in clinical psychology from Virginia Tech in 1989, after completing his predoctoral internship at Children's National Medical Center, George Washington University Medical School, in Washington, D.C. Before he joined Mass General, he was Visiting Assistant Professor on the clinical psychology faculty at Virginia Tech and Assistant Professor in Psychiatry and Pediatrics at the University of Massachusetts Medical Center.

He lives outside Boston with his wife and daughter.

Additional Resources and Support

A variety of resources are available to provide additional information on various topics covered in this book. Here are some of my favorites:

About Parenting Difficult Children

Turecki, Stanley. *The Difficult Child*. New York: Bantam, 1989.

Gordon, Thomas. *P.E.T.: Parent Effectiveness Training*. New York: Plume/Penguin, 1975.

About Problem-Solving Training

Shure, Myrna B. *Raising a Thinking Child*. New York: Pocket Books, 1994.

About Sensory Integration Dysfunction

Kranowitz, Carol Stock. *The Out-of-Sync Child: Recognizing and Coping with Sensory Integration Dysfunction.* New York: Perigee Publishing, 1998.

About Social Skills Deficits

Nowicki, Stephen, and Duke, Marshall. *Helping the Child Who Doesn't Fit In.* Atlanta, GA: Peachtree Publishers, 1992.

About Sibling Issues

Faber, Adele, and Mazlish, Elaine. *Siblings Without Rivalry: How to Help Your Children Live Together So You Can Live Too.* New York: Avon Books, 1998.

About Prosocial Classrooms

Brooks, Robert. *The Self-Esteem Teacher.* Circle Pines, MN: American Guidance Service, 1991.

Charney, Ruth Sidney. *Teaching Children to Care: Management in the Responsive Classroom.* Greenfield, MA: Northeast Foundation for Children, 1992.

Levin, James, and Shanken-Kaye, John M. *The Self-Control Classroom: Understanding and Managing the Disruptive Behavior of All Students, Including Those with ADHD* Dubuque, IA: Kendall/Hunt Publishing, 1996.

Richardson, Rita C. *Connecting with Others: Lessons for Teaching Social and Emotional Competence.* Champaign, IL: Research Press, 1996.

About Nonverbal Learning Disabilities

Rourke, Byron P. *Nonverbal Learning Disabilities: The Syndrome and the Model.* New York: Guilford Press, 1989.

About Deficits in Executive Functions

Lyon, G. Reid, and Krasnegor, Norman A. *Attention, Memory, and Executive Function.* Baltimore: Paul H. Brookes, 1996.

About Children with Tourette's Disorder

Haerle, Tracy. *Children with Tourette Syndrome: A Parents' Guide.* Rockville, MD: Woodbine House, 1992.

About Psychopharmacology

Brown, Ronald T., and Sawyer, Michael G. *Medications for School-Age Children: Effects on Learning and Behavior.* New York: Guilford Press, 1998.

Green, W. H. *Child and Adolescent Clinical Psychopharmacology*. Baltimore: Williams & Wilkins, 1995.

Koplewicz, Harold W. *It's Nobody's Fault: New Hope and Help for Difficult Children and Their Parents.* New York: Times Books, 1996.

Wilens, Timothy E. *Straight Talk About Psychiatric Drugs for Kids.* New York: Guilford Press, 1998.

About Adults with ADHD

Hallowell, Edward M., and Ratey, John J. *Driven to Distraction: Recognizing and Coping with Attention Deficit Disorder from Childhood Through Adulthood.* New York: Pantheon, 1994.

About Legal Issues for Children with Disabilities

Latham, Peter S., and Latham, Patricia H. *Attention Deficit Disorder and the Law*, 2nd ed. Washington, DC: JKL Communications, 1997.

Latham, Peter S., and Latham, Patricia H. *Learning Disabilities and the Law.* Washington, DC: JKL Communications, 1993.

There are also a variety of support groups relevant to inflexible-explosive children, including these:

Children with Attention Deficit Disorders (CHADD)
499 NW 70th Avenue, Suite 101
Plantation, FL 33317
800-233-4050
www.chadd.org

Tourette Syndrome Association (TSA)
42-40 Bell Boulevard
Bayside, NY 11361
718-224-2999
800-237-0717
http://tsa.mgh.harvard.edu/

Learning Disabilities Association of America (LDAA)
4156 Library Road
Pittsburgh, PA 15234
412-341-1515
888-300-6710
http://www.ldanatl.org

National Information Center for Children and Youth with Disabilities (NICHCY)
Box 1492
Washington, DC 20013
800-695-0285
http://www.nichy.org

Obsessive Compulsive Foundation
P.O. Box 70
Milford, CT 06460
203-878-5669
http://pages.prodigy.com/a/willen/ofc.htm/

Anxiety Disorders Association of America
11900 Parklawn Drive, Suite 100
Rockville, MD 20852
301-231-9350
www.adaa.org

National Depressive and Manic-Depressive Association
730 North Franklin Street, Suite 501
Chicago, IL 60610
800-826-3632
www.ndmda.org

National Foundation for Depressive Illness
P.O. Box 2257
New York, NY 10116
800-248-4344
www.depression.org

Depression and Related Affective Disorders Association
Meyer 3-181
600 North Wolfe Street
Baltimore, MD 21287-7381
410-955-4647
www.med.jhu.edu/drada/

American Occupational Therapy Association, Inc.
P.O. Box 31220
Bethesda, MD 20824-1220
800-668-8255
http://www.aota.org

American Speech-Language-Hearing Association
10801 Rockville Pike
Rockville, MD 20852
800-638-TALK
http://www.asha.org

Federation of Families for Children's Mental Health
1021 Prince Street
Alexandria, VA 22314-2971
703-684-7710
http://www.fcmh.org

Index

inflexible-explosive
vs
flexible + frustration tolerant

Child psychologist
"The Out-of-Sync Child"
pp 54-55